Green Energy and Technology

Climate change, environmental impact and the limited natural resources urge scientific research and novel technical solutions. The monograph series Green Energy and Technology serves as a publishing platform for scientific and technological approaches to “green”—i.e. environmentally friendly and sustainable—technologies. While a focus lies on energy and power supply, it also covers “green” solutions in industrial engineering and engineering design. Green Energy and Technology addresses researchers, advanced students, technical consultants as well as decision makers in industries and politics. Hence, the level of presentation spans from instructional to highly technical.

Indexed in Scopus.

Indexed in Ei Compendex.

Dimitris A. Chalkias · Elias Stathatos

The Emergence of Agrivoltaics

Current Status, Challenges and Future Opportunities

Springer

Dimitris A. Chalkias
School of Engineering
Department of Electrical and Computer Engineering
University of the Peloponnese
Patras, Greece

Elias Stathatos
School of Engineering
Department of Electrical and Computer Engineering
University of the Peloponnese
Patras, Greece

ISSN 1865-3529 ISSN 1865-3537 (electronic)
Green Energy and Technology
ISBN 978-3-031-48860-3 ISBN 978-3-031-48861-0 (eBook)
https://doi.org/10.1007/978-3-031-48861-0

This Springer imprint is published by the registered company Springer Nature Switzerland AG
The registered company address is: Gewerbestrasse 11, 6330 Cham, Switzerland

Paper in this product is recyclable.

Όποιου του περισσεύει το κουράγιο και η υπομονή να ερευνά σε όλη του τη ζωή το σκοτάδι, πρώτος θα δει μέσα σε αυτό μια ακτίνα φωτός.

—(D. Glukhovsky) In lieu of dedication.

Preface

The increasing demand for water, energy and food, due to population growth and urbanization, is aggravated by unprecedented extreme weather and climatic conditions. This situation is most likely to undermine the sustainable and peaceful development of humanity in the next years. Today, more than ever, there is an imperative need to support the identification and development of practical solutions, where the use of nexus strategic plans can lead to improved outcomes in the integrated management of water-energy-food-ecosystems (WEFE) resources. Under the aforementioned considerations, the development of agrivoltaic energy systems is nowadays emerging as a very promising cross-sectoral nexus approach that can provide mutual benefits toward the sustainable management of WEFE resources and, therefore, give a quite significant untapped potential for the sustainable development of humanity. Agrivoltaics are hybrids of co-located solar photovoltaic and agriculture infrastructures that can bolster the resilience of renewable energy and food production security, where crops are grown in the partial shade of the solar infrastructure. These energy- and food-generating ecosystems will become a very important—but as yet insufficiently investigated—mechanism for maximizing crop yields, generating renewable energy and efficiently delivering water to plants. The present book contributes to the dissemination of the current knowledge on the modern approach of agrivoltaics, providing a comprehensive state of the art on the field, discussing the current status, the challenges and the future perspectives for their further development, from conceptual designs to practical realizations.

Patras, Greece

Dimitris A. Chalkias
Elias Stathatos

Contents

Abbreviations

3BL	Triple bottom line
μc-Si:H	Hydrogenated micro-crystalline silicon
a-Si:H	Hydrogenated amorphous silicon
a-Si_xGe_{1-x}:H	Hydrogenated amorphous silicon germanium
ABTS	2,2′-azino-bis-3-ethylbenzthiazoline-6-sulphonic acid
Al-BSF	Aluminium back surface field
AR	Anti-reflective
AVT	Average visible light transmittance
BOS	Balance of system
c-Si	Crystalline silicon
CAGR	Compound annual growth rate
CdTe	Cadmium telluride
CGF	Crop growth factor
CIGS	Copper indium gallium diselenide
COP26	26th United Nations Climate Change Conference of the Parties
CS	Column spacing
CVD	Chemical vapor deposition
DIN	German Institute for standardization
DPPH	1,1-diphenyl-2-picrylhydrazyl
DSSC	Dye-sensitized solar cell
DSSM	Dye-sensitized solar module
DSSP	Dye-sensitized solar panel
EPBT	Energy payback time
ESMAP	Energy sector management assistance program
EU	European Union
FAO	Food and agriculture organization
GCE	Ground coverage ratio
GDP	Gross domestic product
GHG	Greenhouse gas
HBC	Heterojunction with interdigitated back contact
het-Si:H	Hydrogenated heterogeneous silicon

HIT	Heterojunction with intrinsic thin-layer
HJ	Heterojunction
HOMO	Highest occupied molecular orbital
IBC	Interdigitated back contact
ICS	Inter-column-spacing
IEA	International Energy Agency
IRS	Inter-row-spacing
ITRPV	International Technology Roadmap for Photovoltaic
LCA	Life cycle assessment
LCI	Life cycle inventory
LCIA	Life cycle impact assessment
LCOE	Levelized cost of electricity
LED	Light-emitting diode
LER	Land equivalent ratio
LSC	Luminescent solar concentrator
LSPR	Localized surface plasmon resonance
LUMO	Lowest unoccupied molecular orbital
mc-Si	Multi-crystalline silicon
MJ	Multi-junction
Mtoe	Megatonnes of oil equivalent
MZO	Magnesium zinc oxide
na-Si:H	Hydrogenated nano-amorphous silicon
nc-Si:H	Hydrogenated nano-crystalline silicon
nc-SiO_x:H	Hydrogenated nano-crystalline silicon oxide
OPV	Organic photovoltaic
PAR	Photosynthetic active radiation
PCE	Power conversion efficiency
pc-Si:H	Hydrogenated proto-crystalline silicon
PEN	Polyethylene naphthalate
PERC	Passivated emitter rear contact
PERL	Passivated emitter rear locally diffused
PERT	Passivated emitter rear totally diffused
PSC	Perovskite solar cell
PET	Polyethylene terephthalate
pm-Si:H	Hydrogenated polymorphous silicon
poly-Si	Poly-crystalline-silicon
PV	Photovoltaic
PV_R	Photovoltaic coverage
PVD	Physical vapor deposition
QD	Quantum dot
QDSC	Quantum dot solar cell
RS	Row spacing
sc-Si	Single-crystalline silicon
SDG	Sustainable development goal
SSC	Spectral splitting covering

TCO	Transparent conductive oxide
T_g	Glass transition temperature
TOPCon	Tunnel oxide passivated contact
WEFE	Water, energy, food and ecosystem
WUE	Water usage efficiency

Chapter 1
The Water-Energy-Food-Ecosystems (WEFE) Nexus

This chapter is the forerunner of the cross-sectoral nexus approach of agrivoltaics that the present book introduces, presenting the current challenges on the Water-Energy-Food-Ecosystems (WEFE) nexus, as well as approaches that should be adapted in order to attain sustainable management of the WEFE nexus.

1.1 The Challenge

Water, energy, food and ecosystems (WEFE) are four terms that, at first glance, appear to be completely unconnected but actually are highly related to each other, and their correlation is highly important and pretty essential for the sustainable development and prosperity of humankind [1]. Most projections show that the demand for freshwater, food and energy will highly increase over the next decades under vague population growth, as well as due to cultural and technological changes. Simultaneously, the ever-increasing greenhouse gas emissions will lead to global warming, climate change and, subsequently, to ecosystems disruption. So, this is the time more than ever that humanity should adopt effective strategies toward sustainability. To this aim, the "WEFE nexus" is considered the ideal approach to the sustainable management and governance across the multiple sectors of WEFE. This is because the cross-sectional nexus approach ("nexus" term was originally derived from the Latin verb "nectere", which means "to connect") denotes any interaction among water, energy, food and ecosystems independent sectors, taking into account any appeared synergies and conflicts that arise from the way they are managed (see Fig. 1.1) [2].

The WEFE nexus provides a foundation for a system of systems that connects the WEFE subsystems. It complements, rather than replaces, existing disciplines. Interlinkages, hotspots and tradeoffs are nexus components. Nexus is based on the water productivity concept familiar in agriculture, as well as on integrated water resource management, energy efficiency and other aspects. It links government

D. A. Chalkias and E. Stathatos, *The Emergence of Agrivoltaics*,
Green Energy and Technology, https://doi.org/10.1007/978-3-031-48861-0_1

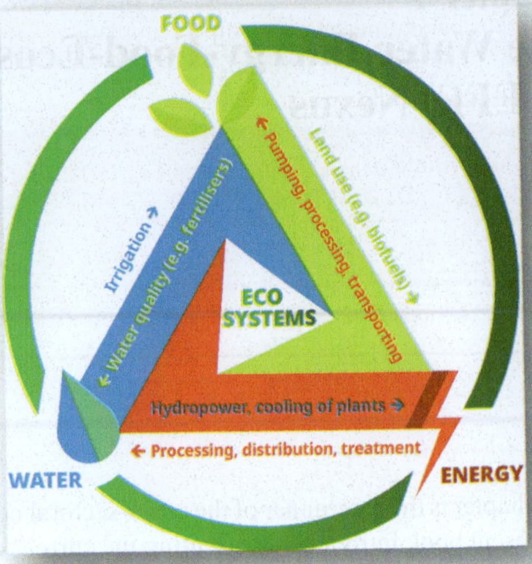

Fig. 1.1 The water-energy-food-ecosystems nexus [3]

policy, society and commercial supply networks. The WEFE nexus does not end there, but goes beyond analytics to provide a framework or platform that connects science to economy and politics. It is the platform that moves the balance from conflict to cooperation [4].

Building the WEFE nexus system entails: (1) defining the food supply chain components, (2) establishing the water, food and energy resource portfolios, (3) estimating the resource footprint, (4) identifying the interventions that need to be studied, (5) developing the tool and performing tradeoff analysis for various scenarios, (6) the sustainability index and (7) the dialogue. Finally, two forms of discussion are required: one with only the scientific community and another with the broader society, i.e., scientists, decision-makers and others engaged in the resource nexus [4].

To better understand the interconnection of the WEFE subsystems and the subsequent challenges that humanity faces, the following discussion is given as an example. Water used in the agricultural sector accounts for 70% of total global freshwater consumption, making it the largest user of water [5]. Energy is required to extract, pump, collect, transport and treat water, and finally, transport and distribute food as well. Along with water and energy, food production and supply chain consume about 30% of the total energy consumed globally [6]. Industry, residential around the world and finally, any end-user claim more energy, water and land resources for farming that may cause environmental degradation and scarcity of the basic resources. This phenomenon is expected to be sharper in the very near future as approximately 60% more food will be required to cover the world population in the next 30 years [7]. As this demand is gradually growing, there will also be an increase in competition for resources between energy, agriculture, forestry, water, fisheries, transport and other sectors, with unpredictable effects on the environment and livelihood in general. On the other hand, energy consumption is continuously growing year by year, and it is believed that in the next three decades, the need for energy will be 50% higher [8]. Besides, the need for water, either for farming or for irrigation, is also believed to increase by 20–30% by 2050 [9]. Finally, the constant search for resources, such as water, energy and food, can create various social effects but, most importantly, can bring changes in ecological systems.

General Secretary of the United Nations, in his talk on "World Water Day" in 2011, noticed that the specific interconnection between water, energy, food and ecosystems is among the greatest challenges that mankind faces today. After that, a conference focused on "Water, Energy and Food Security Nexus: Solutions for the Green Economy" was announced later that year in Bonn [10]. European Commission also agrees with previous concerns, while Policy Officer at the Directorate-General Research and Innovation emphasizes that "it is tremendously important to transit to a more sustainable and integrated resource management, respecting the specific interlinkages and interdependencies of these individual sectors". From the European Commission's Joint Research Centre, Giovanni Bidoglio adds: "By matching the interests of different stakeholders and economic sectors, nexus offers opportunities for the design of shared solutions to control conflicts, achieve sustainability and reduce inequalities". The Managing Director of the Union of the Mediterranean, Almotaz Abadi, adds that "apart from resource scarcity, the Mediterranean has to deal with conflict, migration and all challenges are closely intertwined". He explains: "In the Mediterranean, more than 85% of water is used for agricultural purposes, and the region also has the highest dependency on food imports. Over the years, agricultural productivity has not increased in a meaningful way. Food loss and waste have reached alarming levels with huge implications on water, land, energy and greenhouse effect and gas emissions. The region is a global hotspot for the over-exploitation of groundwater, a resource that will be rapidly depleted in the coming decades. In light of this, future agricultural policies will heavily need to rely on other irrigation sources, such as wastewater reuse, while following more stringent guidelines for reducing waste. Opportunities around demand management of water are well articulated but underutilized. Interlinkages across sectors (water, food, energy) and the natural resources base (water, land, soils) are understood but not adequately reflected in policies" [11].

Due to the aforementioned challenges, strategies based on the WEFE nexus have been proposed as an interesting combination for the effective treatment of complex problems related to natural resources but also the need for energy to meet environmental goals, the economy and social groups (see Fig. 1.2). Nevertheless, any experience until now has shown that the conceptualization and application of WEFE nexus strategies in practice is not easy and, in many cases, has a lot of ambiguity [12]. Even the term "nexus" has a number of meanings. According to the Oxford English Dictionary, the term "nexus" is defined as (1) a mean of connection between parts (a link or bond), (2) a connected group (a network), (3) a point of convergence (a focus). In the context of the WEFE, the first definition emphasizes the interconnectedness of the subsystems of water, energy, food and ecosystems, while the second emphasizes each subsystem membership in the cluster. The third definition, on the

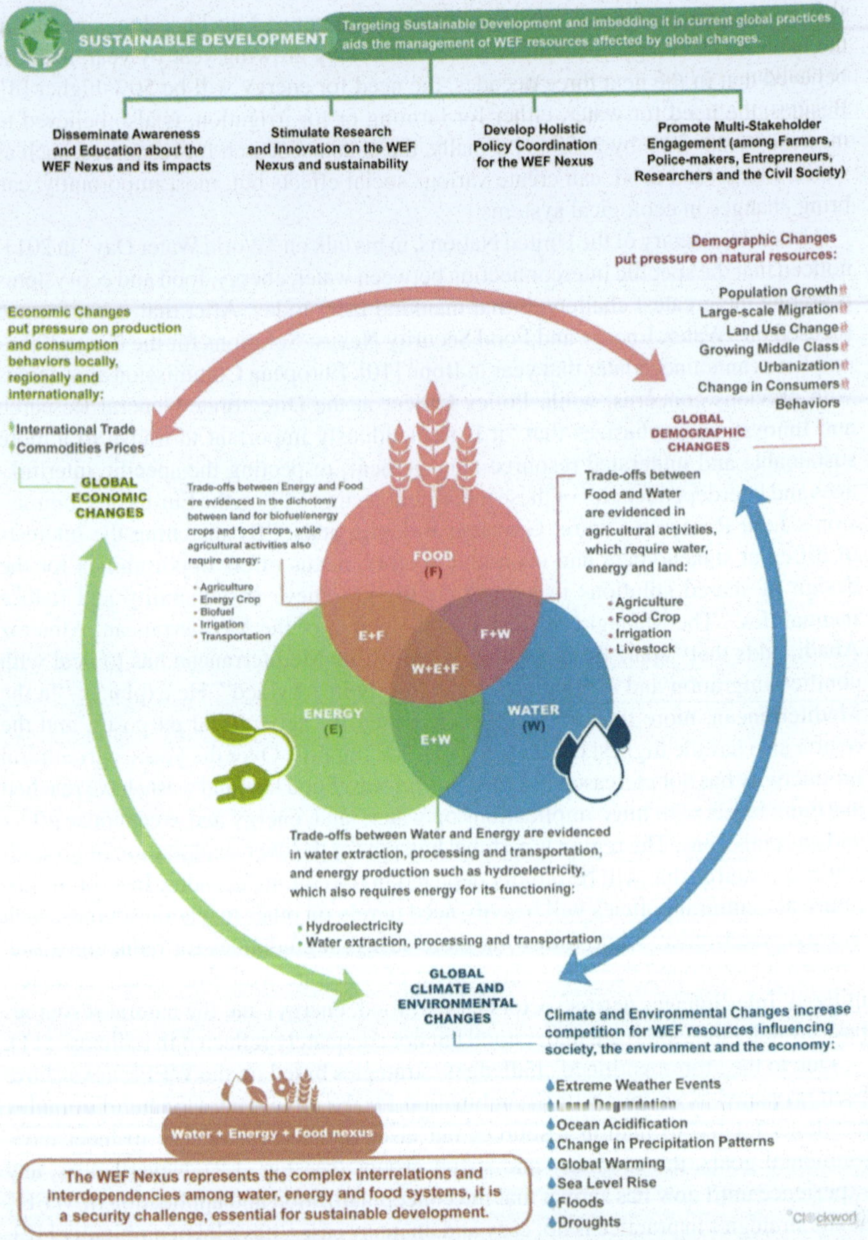

Fig. 1.2 The water-energy-food nexus for sustainable development [13]

other hand, refers to their merger with the social environment in which they have significance. The ambiguity and vagueness associated with the WEFE nexus raise the challenge of how to make sense of the complexity of the approach in order to take the appropriate actions. Herein, two major issues have to be solved: sustainability and governance. This necessitates an integrated approach to thinking about these issues, evoking "systems thinking" [12].

1.2 Perspectives for a Sustainable Water-Energy-Food-Ecosystems Nexus

There are many uncertainties on how societies can effectively adapt to global sociodemographic, economic and environmental changes. At the core of the interest is the need to promote the sustainable management of the WEFE nexus, considering both natural and anthropogenic factors. In this regard, the sustainable development goals (SDGs) of the United Nations (see Fig. 1.3) aim to support governments and decision-makers in their attempts to improve resource management efficiency, transparency and accountability, while also identifying additional and alternative resources to promote sustainable development. Promoting the holistic and multi-sectoral management of the WEFE challenges will require government assistance, education and awareness, cross-sectoral and multi-level collaboration among stakeholders, research and technical development, innovations and new manufacturing methods [13].

In order to attain sustainable management of the WEFE nexus, key recommendations should be taken into account. These include: improved natural resource management under the continuous technological development and adoption of approaches to increase the resilience of critical infrastructures; promotion of the engagement of multiple stakeholders (scientists, business persons, governments and civil society) to increase the synergies of the WEFE nexus, instead of encouraging negative externalities/trade-offs; probabilistic analysis and stochastic models for obtaining reliable data to aid risk management decision-making (the WEFE nexus comprises complex challenges and diverse hazards that need stochastic analysis in order to minimize uncertainties and enhance multi-stakeholder participation for cross-sectoral coordination and policy-making); increased investment in research and development to attain breakthroughs in science, technologies and processes; increase economic sector interconnectedness and policy cooperation to minimize unforeseen consequences [13].

Table 1.1 tabulates some practices that are recommended to support the sustainable management of the WEFE nexus. The presented recommendations are related to the SDGs.

Fig. 1.3 Sustainable development goals set by the United Nations Foundation [14]. (Statement: The content of this publication has not been approved by the United Nations and does not reflect the views of the United Nations or its officials or Member States)

Table 1.1 Practices employed to support the sustainable management of the water-energy-food-ecosystems nexus [13]

Practice	Benefits	SDGs
Reduction of post-harvest losses using efficient harvesting, transport and cooling systems	Increase in WEF resource efficiency by reducing wastes	12, 13
Usage of organic waste to produce bioenergy	Reduction of the competitiveness between the land usage for food and biofuel production, while also reduction of the levels of waste	7, 9, 11, 12, 13
Utilization of technology to increase the efficiency of water use, including conservation agriculture, micro-irrigation and rainwater storage	Increment of water security and improved access to clean water for all	2, 6, 9, 11, 12, 13

(continued)

Table 1.1 (continued)

Practice	Benefits	SDGs
Cultivation of vertical crops in urban areas for increased local fresh food production, energy saving, and cost of packaging and transportation reduction	Optimization of the urban areas usage and reduction of competition for land. Aligned with other technologies, such as rainwater harvesting, sewage treatment and biogas production, vertical crops increase resilience to the WEFE Nexus and its complexities	2, 9, 11, 12, 13
Diversification of food production, expansion of mariculture, production of seaweeds and marine animals	Reduction of the pressure on agribusiness for the production of food	2, 9, 11, 12, 13, 14, 15
Development and implementation of agrivoltaic systems	Improved urban areas usage and reduction of competition for land. Increased energy and food production	2, 7, 9, 11, 12, 13, 15
Implementation of pilot projects as case studies to create innovative solutions to the WEFE challenges	Pilot projects provide real data on the WEFE Nexus, and their adoption should be encouraged for all sectors to increase knowledge regarding the complexities of the Nexus and its synergies and trade-offs in multiple sectors	9, 11, 13, 16, 17
Promotion of capacity building and training on the holistic/multi-sectoral and multi-disciplinary approach to the WEFE nexus	Educate stakeholders regarding the impacts of climate change on WEFE security through knowledge sharing and capacity building	4, 12
Promotion of multi-stakeholder engagement in communities of practice for the WEFE nexus	Development of communities of practice for the WEF Nexus can provide information on local/regional challenges, and provide multi-sectoral, multidisciplinary and multi-stakeholders solutions to these challenges	9, 11, 16, 17

1.3 Outlook

Achieving water, energy, food and ecosystems security is one of the greatest challenges facing humankind today. To this aim, the WEFE nexus has emerged as a multi-centric lens for assessing their integrated sustainable management. As a fresh way of thinking, WEFE nexus strategies and practices are nowadays increasingly under investigation, application and evaluation to support sustainable development and deal with the failures of sector-driven management strategies. Today, a response to the call for a WEFE nexus approach that is systemic, can manage its associated complexity, ambiguity and vagueness, as well as the multiple stakeholders, each with their subjective viewpoint, and the implied governance implications, is a challenging but at the same time imperative step towards sustainability. The call to operationalize the nexus is increasingly heralded, and a common theme is that 'nexus thinking' must evolve into 'nexus action'. To this end, this book aims to present the strategic development of "agrivoltaics" as an emerging and very promising approach that can enable and achieve 'nexus doing', providing mutual benefits across the WEFE nexus.

References

1. Salam PA, Shrestha S, Pandey VP, Anal AK (2017) Water-energy-food nexus: principles and practices, first. Wiley
2. De Laurentiis V, Hunt DVL, Rogers CDF (2016) Overcoming food security challenges within an energy/water/food nexus (EWFN) approach. Sustainability 8:95
3. Global Water Partnership-Mediterranean (2020) Water-energy-food-ecosystems nexus. https://www.gwp.org/en/GWP-Mediterranean/WE-ACT/Programmes-per-theme/Water-Food-Energy-Nexus/. Accessed 20 Nov 2022
4. Mohtar RH (2022) The WEF nexus journey. Front Sustain Food Syst 6:820305
5. Food and Agriculture Organization of the United Nations (2017) Water for sustainable food and agriculture
6. IRENA and FAO (2021) Renewable energy for agri-food systems: towards the Sustainable Development Goals and the Paris Agreement. Abu Dhabi and Rome
7. Da Silva JG (2012) Feeding the World Sustainably. In: United Nations. https://www.un.org/en/chronicle/article/feeding-world-sustainably#:~:text=According to estimates compiled by, toll on our natural resources. Accessed 20 Nov 2022
8. Energy Information Administration (EIA) (2021) International Energy Outlook 2021
9. Boretti A, Rosa L (2019) Reassessing the projections of the World Water Development Report. npj Clean Water 2:15
10. Bonn2011 Nexus Conference (2011) The water, energy and food security Nexus—solutions for a green economy. https://www.water-energy-food.org/events/conference-the-water-energy-and-food-security-nexus-bonn2011-nexus-conference. Accessed 20 Nov 2022
11. Partnership for Research and Innovation in the Mediterranean Area (PRIMA) (2022) Green Mediterranean. https://prima-med.org/wefe-nexus-community-of-practice-a-solution-for-the-mediterranean-future/. Accessed 20 Nov 2022
12. Harwood SA (2018) In search of a (WEF) nexus approach. Environ Sci Policy 83:79–85
13. de Andrade GJBSO, Berchin II, Garcia J et al (2021) A literature-based study on the water–energy–food nexus for sustainable development. Stoch Environ Res Risk Assess 35:95–116
14. United Nations 17 Goals to Transform Our World. https://www.un.org/sustainabledevelopment/. Accessed 12 Jul 2023

Chapter 2
Energy and Agriculture

This chapter, entitled "Energy and Agriculture", accounts for the energy demands of the agricultural sector, discussing, in brief, the main energy inputs at all stages of modern agricultural production and how they are usually managed up until today. The focus herein is to demonstrate the emerging eco-agricultural trend, presenting possible sustainable solutions for energy efficiency and conservation, as well as the ever-increasing application of renewable energy sources to meet the arising energy demands. The aim is to raise awareness of the readers towards the establishment of the mentality of sustainable agriculture that can strike a balance between maximizing crop productivity, preserving economic stability and minimizing the usage of finite natural resources and the detrimental impacts on the environment.

2.1 Introduction—Towards a Sustainable Agriculture

The agricultural sector has changed dramatically in the last century. Food and fiber productivity that previously depended only on natural energy flows from the sun and chemical energy stored biologically in the soils has now soared due to mechanization, specialization, new technologies, increased chemical usage and government policies that favor the maximizing of production and reduction of food prices [1]. Additional energy inputs are increasingly given by humans' interference to make a given area of land or unit of water produce more food than it would be done naturally. However, even under this rapid pace of development, the need for the race for further yield improvements is continuous to match the growing demands of humanity. It is expected that, by 2050, the global demand for energy will rise by 60%, water withdrawals by 55% and food demand by 60% as a result of the continued population and economic growth [2].

Meeting, at least in part, the growing global food demand over the past century has been achieved by a significant increase of fossil fuel inputs along the whole

D. A. Chalkias and E. Stathatos, *The Emergence of Agrivoltaics*,
Green Energy and Technology, https://doi.org/10.1007/978-3-031-48861-0_2

agri-food chain. However, this high dependency of the food system on fossil fuels is now becoming a source of worry. These increasing demands are met in the context of an already stressed natural resource asset, limited availability of productive land area and climate change impacts. Agriculture and the entire agri-food supply chain today contribute approximately 22% of total annual anthropogenic greenhouse gas emissions, plus an additional 15% contributed from land-use changes, including deforestation to gain more agricultural land area [3]. This environmental destruction, combined with topsoil depletion, lurk global prosperity and sustainable development. Analyses of the probable impacts of climate change on agricultural productivity up to 2050 have shown that negative effects could lead to a reduction in food availability and human well-being, particularly in developing regions [3]. Furthermore, under this trend, the problem also acquires social dimensions since there is also disintegration of small rural communities, with family farms declining, while the working and living conditions of farm laborers are neglected [1].

During the past last decades, there has been a growing movement to question the necessity of this kind of agricultural development and to offer innovative alternatives. The global economy seeks sustainable solutions to make a major transition from business-as-usual to address these challenges. Today, sustainable agriculture is garnering increasing support and acceptance within food production systems. Sustainable agriculture is in line with the "triple bottom line" (3BL) framework, integrating three main goals: environmental stewardship, economic profitability and social responsibility (Fig. 2.1) [4]. This differs from a solely profit-driven approach, where businesses prosper monetarily at the cost of the environment and society. In line with this emerging eco-agricultural trend, a sustainable agricultural system should be characterized by enhanced energy efficiency and conservation, while it cannot depend on exhaustible (finite) resources such as fossil fuels, but it should be based on the prudent usage of renewable and/or recyclable resources.

2.2 Energy Demands in Agriculture

The modern agricultural sector requires a broad collection of energy inputs. Energy is required for all steps along the agri-food chain, such as for production, processing, storage and transport, for preparing fertilizers, agro-chemicals, etc., while significant are the amounts of energy (about 30%) that are lost through food losses. According to the Food and Agriculture Organization (FAO), the agri-food sector currently accounts for around one-third of the world's total energy consumption [6]. The energy input for the agri-food sector to the total end-use energy differs from nation to nation. In countries with a high gross domestic product (GDP), about 25% of the total agri-food energy demands regard production at the stage "behind the farm gate", 45% for food processing and distribution, and 30% for retail, preparation and cooking (see Fig. 2.2). In countries with a low GDP, a smaller share of energy is used on-farm and a higher share for preparation and cooking. By considering the aforementioned, it is perceived that only a small part of total food energy demands regards primary farm

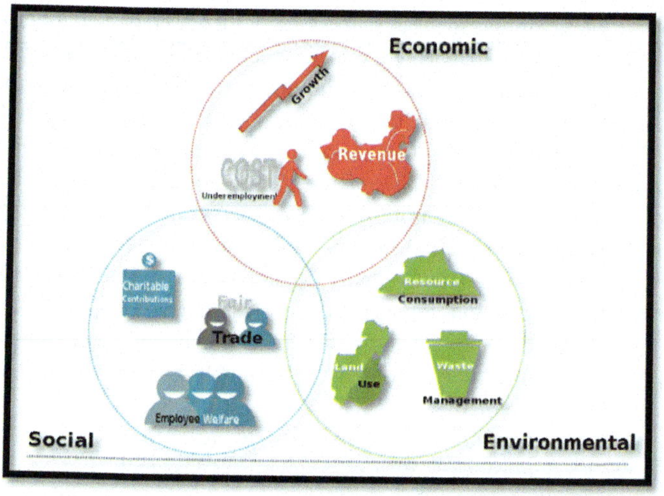

Fig. 2.1 Schematic representation of the triple bottom line framework [5]. (Reproduced from Ref. [5], creator: clonewayx, under the terms of the CC BY-SA 4.0 license)

production. However, according to several studies, this part of the agri-food chain produces about 65% of the total food sector's greenhouse gas emissions and should be seriously considered to attain sustainability [3].

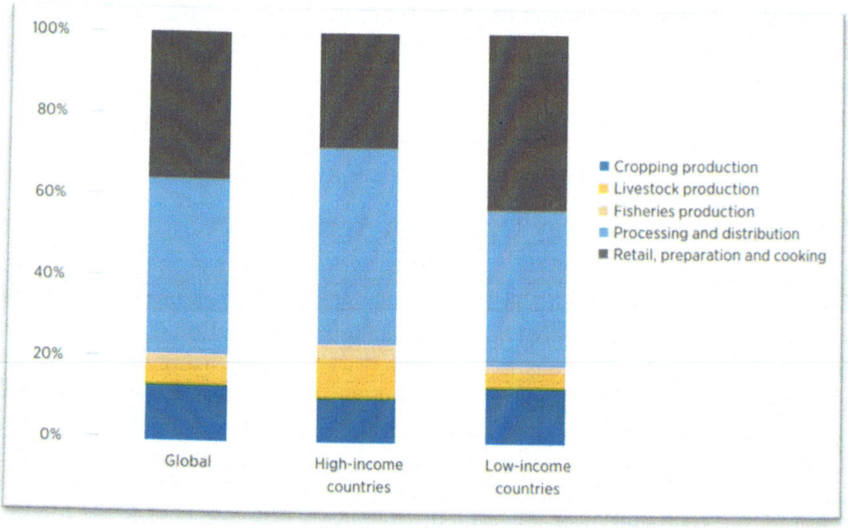

Fig. 2.2 Energy share by segment of agri-food chain globally and in high- and low-income countries [7]. (Reproduced with permission from FAO)

Worldwide, energy inputs in the agri-food sector increased by >20% between 2000 and 2018. A key reason for that growth was the mechanization of Asian agriculture, in the form of farm machinery, processing equipment, irrigation pumps and inputs, e.g., fertilizers. In America and Europe, the situation is considered stable, even though the production has increased, thanks to enhanced technological and agronomic progress. In Africa, which hosts around 15% of the global population and faces growing demand for food, energy consumption has remained constant, accounting for only 4% of global energy consumption in agri-food systems [7]. In the last decade, the purely agricultural sector consumed approximately 1700 trillion Btu of energy each year [8]. For European Union countries, the corresponding energy consumption in agriculture is estimated at a mean value of 3.3% of final energy consumption (2019 statistics), with the highest shares found in the Netherlands (8.9%) and Poland (5.5%) [9].

Agri-food energy demands can be divided into on-farm and off-farm. Farming and agricultural production, as well as occasional and seasonal labor, are included in on-farm activities. On the other hand, off-farm encompasses all agriculture-related activities that take place beyond the farm. Viewed through a value chain lens, off-farm includes the "middle" and "end" of the process by the time agricultural goods leave the farm to ultimately reach consumers [10]. At the farm level, energy use is classified as either direct or indirect (see Table 2.1). Direct energy use in agriculture is primarily related to fuels or electricity to operate machinery and equipment for preparing fields, planting, cultivation and harvesting crops, post-harvest processing, irrigation, application of chemicals, and transporting inputs and outputs from and to the market. Furthermore, energy needs exist for lighting, heating and cooling. On the other hand, indirect energy needs are in the form of sequestered energy in fertilizers, pesticides, herbicides and insecticides. Across each of these activities, there are major variations in energy needs, depending on the scale of the enterprise and the food type being produced.

2.3 Sustainable Energy Options in Agriculture

The sustainable provision of the increasing amounts of energy and water demands in the agricultural sector is essential in a world facing both population growth and rising nutritional challenges. This has to be achieved without increasing greenhouse gas emissions. Such a scenario requires the development and implementation of innovative integrated agricultural approaches. Several sustainable technologies suitable for agriculture applications already exist in large numbers and at commercial scales, but this does not take away the need to continuously develop new technologies. However, the rate at which these technologies are adopted and further developed is dependent on favorable political, policy and financial conditions. This requires close cooperation between policymakers, agricultural and financial sectors.

Table 2.1 Energy usage in agricultural production [11]

Direct use of energy	*Source of energy*
Operating machinery and trucks: – Fieldwork (tractors, combine harvesters, mowers, etc.) – Input purchase and deliveries (trucks)	Diesel fuel
Operating equipment: – Irrigation equipment – Drying of grain or fruit – Standby generators – Heating/cooling of facilities, greenhouses, stock tanks, etc. – Animal waste treatment	Diesel fuel Natural gas Liquid propane gas Electricity
General farm overhead – Lighting for houses, sheds and barns – Power for farm household appliances	Electricity
Marketing – Transportation: elevator to terminal, processor, or port – Elevating	Diesel fuel Gasoline
Indirect use of energy	*Source of energy*
Fertilizer – Nitrogen-based (Natural gas is 75–90% of the cost of prod.) – Phosphate (Natural gas is 15–30% of the cost of prod.) – Potash (Natural gas is about 15% of the cost of prod.)	Natural gas
Pesticides (herbicides, insecticides, fungicides)	Petroleum or natural gas

2.3.1 Enhancing Energy Efficiency and Conservation

The first crucial step that should be taken toward the achievement of more sustainable agriculture is to enhance energy efficiency and conservation [3, 12]. Improving energy efficiency and promoting energy saving in the agricultural sector are essential ways to reduce energy demands and, subsequently, food costs. Thus, greenhouse gas emissions will also be reduced, tackling climate change.

Initially, to achieve energy efficiency and conservation, the areas of major energy consumption in the under-consideration agricultural system should be defined. Heating, lighting, ventilation and refrigeration equipment are generally the major energy consumers, and therefore, with their proper management, a significant enhancement in energy efficiency and saving can be realized (see Table 2.2). In addition, animal feeding and waste removal systems may have a significant energy footprint. To identify areas of high energy consumption, farmers must first benchmark energy usage. This involves the collection of information on the amount of energy used by their activities. The energy consumption can be compared against similar farms and industry averages to indicate the energy performance of their farming operation.

There are several very low-cost options to improve the energy efficiency of agricultural activities. Switching off approach is the main cost-free opportunity for energy saving. When not in use, equipment should be turned off. Alternatively, time switches

Table 2.2 Areas of high energy consumption depending on farming activity [12]

Farming activity	Areas of high energy consumption	Fuel type
Field horticulture	– Field operations: tractors, crop treatment machinery – Crop storage	Diesel fuel Electricity
Indoor horticulture	– Heating of greenhouses – Lighting of greenhouses	Natural gas Coal Liquid propane gas Electricity
Arable crops	– Grain drying, storage and handling	Electricity
Potatoes and sugar beet	– Field operations – Crop preparation – Crop storage	Diesel fuel Electricity
Intensive livestock (e.g., pigs, poultry and eggs)	– Maintenance of environmental conditions for stock rearing	Electricity
Beef and sheep	– Forage production	Diesel fuel Electricity
Dairy	– Water heating – Refrigeration of milk – Milking machinery	Diesel fuel Electricity

can significantly reduce energy usage. Maintenance of the machinery is also important. For instance, boilers maintenance, keeping machinery components clean and checking of the seals are all measures that can easily lead to energy conservation. Careful selection of equipment is also important during the procurement process. It is important to test whether the equipment will be able to meet the system demands that are intended to be placed into. Correct power-sizing of equipment (e.g., motors) should be considered to optimize energy efficiency. When also combined with a rigorous maintenance system, more efficient functioning and a longer lifetime can be achieved. On the other hand, equipment procurement should not be entirely based on upfront cost because, choosing more energy-efficient equipment can be more beneficial to the farming operation overall (considering its lifetime). Life cycle costing is important to evaluate since it takes into account annual operating costs, which will be lower if more efficient equipment is procured. Furthermore, control systems are a powerful tool for assessing energy usage. Many modern control systems can store temperature and ventilation data, which can be used to manage energy inputs. Utilizing control systems to manage operating conditions can significantly increase the energy efficiency of agricultural systems without compromising food production.

Heating is often a highly energy-consuming area, which gives a significant opportunity for energy efficiency and conservation. Heating is a key part for the maintenance of many agricultural facilities and should be ensured that it matches the requirements of the activities. Initially, heating control systems should be located in the correct areas, in terms that are not affected by side effects. Furthermore, by the correct maintenance of the boilers and pipework, heating costs can be reduced. The use of modern types of heaters, such as radiant heaters, can also lead to more efficient heating, allowing heat to be directly targeted at a specific building sector. The use of compensators is also considered an upgrade of the temperature control system since they can regulate heating depending on the weather and/or forecast the time until reaching optimum environmental conditions, subsequently minimizing energy losses. Finally, it should be considered that agricultural and horticultural buildings vary widely in terms of build quality, age and type. Identifying the opportunities for enhancing the energy efficiency of buildings (insulation) is quite important since it is responsible for even more than 25% of building heat losses.

Refrigeration is also applied in many sections of the agricultural industries, such as dairy farming, and its implementation can comprise a significant proportion of energy costs. Maintenance of refrigerators in terms of keeping clean the condensers is important since it can lead to significant energy saving. Upgrading self-closing doors can also significantly reduce temperature losses, while avoidance of overcooling can increase the energy efficiency of these systems. Finally, it is important to ensure that refrigerators are not overloaded to ensure their efficient operation.

Ventilation is often an underappreciated area of energy consumption in agricultural systems. Ventilation accounts for about 14% of energy costs across the sector. Regular ventilation system maintenance is important for energy efficiency. The effectiveness of ventilation systems can fall by up to 60% if not properly maintained, leading to increased running costs and risk of breakdown. Natural ventilation of the agricultural facilities leads to energy saving. In many cases, windows are able to provide sufficient levels of natural ventilation, which can allow fans to be powered down or even switched off. However, under this approach, wider implications, such as the impact on heating demands and security, should also be considered. Identifying and implementing energy savings opportunities in ventilation systems can have a significant impact with a possible reduction of energy consumption of up to 20%. Control of the speed of ventilation systems allows the matching of the requirements, leading to energy savings. Automatic control utilizing time switches and sensors can reduce the operating time of fans and, subsequently, energy consumption.

Lighting accounts for about 4% of energy costs across the agricultural sector, and therefore, having an efficient approach to lighting usage can lead to energy saving. Replacing conventional tungsten or compact fluorescent light bulbs with light-emitting diodes (LEDs) can decrease energy demands for lighting and provide better light quality. Additionally, the installation of sensors that can ensure that lighting is only used when required is important. Daylight sensors can be used to dim or switch off lighting when natural daylight is sufficient, and it is particularly effective for areas that are intermittently used. This can lead to a saving on lighting energy costs of 30–50%.

The path to the achievement of sustainable agriculture (in the long run) is also related to on-farm technological progress. Principal strategies that today's scientific community investigates include the development of more efficient machinery, improved agrochemical management (e.g., precision farming with individualized management of pesticides, nutrients and water, as well as computer-controlled livestock feeding), more efficient irrigation systems, more efficient heaters, improved insulation, more efficient light sources, even by plant and livestock genetic improvements.

2.3.2 Renewable Energy Sources in Agriculture

In our days, renewable energy gains high momentum in the agricultural sector under the rapid development of new technologies and continues falling costs [13]. Renewable energy sources can now provide a competitive energy supply with lower risks, contributing simultaneously to the fast and secure independence from fossil fuel imports. Today, renewable energy and farming are considered a winning combination. Biomass, solar, wind, hydropower and geothermal energy can be harvested forever, providing farmers with a long-term source of income. At first, they can be used on the farm to replace traditional fuels such as coal, natural gas and oil. On the other hand, the energy produced by renewable energy sources can be sold as a "cash crop", such as selling the produced electric power to the grid, providing an extra income to the farmers. Yet, several challenges are imposed under this approach, mainly related to a possible intermittent energy supply, the highly regionally distributed nature of renewable energy sources and the high initial costs [13, 14].

Bio-energy Strategies in Agriculture

Recently, the bio-energy sector has seized increasing attention as an imperative countermeasure for a low-carbon environmental future. Based on the International Energy Agency (IEA), the demand for bio-energy is escalating, and experts have forecasted that it will account for more than 17% of global energy by 2060 [15]. The global bio-energy market size was valued at $344.90 billion in 2019 and is projected to reach $642.71 billion by 2027, exhibiting a compound annual growth rate (CAGR) of 8.0% during the forecast period 2020–2027 [16]. Along with the other renewable energy sources (e.g., solar, wind, hydrothermal and geothermal), bio-energy is receiving increasing attention because bio-energy feedstock is highly available, effective in greenhouse gas emission mitigation, and most importantly, it is hailed as a "win–win" carbon credit offset strategy to eliminate the biomass waste through its conversion to value-added chemicals [15].

Bio-energy is the energy of biological and renewable origin, and refers to the technical systems through which biomass is collected and used as an energy source to produce heat, electricity or transport fuels [17]. Biomass conversion can be achieved through thermochemical conversion processes within a gas atmosphere (e.g., combustion, pyrolysis, carbonization, torrefaction, gasification) or within a hydrothermal atmosphere (hydrothermal liquefaction, hydrothermal carbonization, hydrothermal gasification), through biological or biochemical conversion processes (e.g., fermentation, anaerobic digestion and enzymatic processes), as well as through physicochemical conversion processes (e.g., pressing and chemical treatment) [18]. Globally, there is a wide variety of feedstocks available for bio-energy production, including agricultural and forestry residues, industrial wastes and municipal solid wastes. Amongst them, the management of agricultural wastes to produce useful bio-energy has gained significant attention in recent years toward the development of more sustainable agriculture [19].

Agricultural biomass is an abundant energy source, with the characteristics of renewable, carbon–neutral and wide distribution. The utilization of agricultural biomass resources in current coal-fired plants can not only solve environmental pollution arising from inappropriate treatment but also mitigate the greenhouse gas emissions and energy crisis from fossil fuels scarcity. The approximate fuel value of crop residues at a global level is estimated at $3758 \cdot 10^6$ Mg/year, which has an oil equivalent of $7560 \cdot 10^6$ barrels [20]. Agricultural biomass can be utilized as a source of energy in various forms, depending on its physical and chemical properties, and its availability. Amongst the most important physical properties to be considered are particle size, density, grindability and flowability [21]. The conversion efficiency and energy requirements vary depending on the size and shape of the particle units. In most cases, biomass should be comminuted to convert the particle from as-received condition to desirable sizes for the user. Bulk density plays an important role in energy conversion since it affects its handling, storage and transportation. Grindability is also an important criterion for the selection of the appropriate type of biomass source material for bio-energy production since grinding is an energy-consuming process. The flow behavior of the granular biomass material is also important to be aware since handling issues, such as screw feeder clogging and hopper arching, can highly increase the equipment functioning time. On the other hand, the chemical properties of the biomass determine the bio-energy conversion efficiency. To achieve an efficient bio-energy conversion, proximate analysis, ultimate analysis, heating value analysis and compositional analysis are important to precede [21]. The determination of the fixed carbon content through proximate analysis is important since it gives information on the amount of char formation in the thermochemical conversion processes. The yield of char produced during thermochemical conversion processes increases as fixed carbon concentration increases. On the other hand, ultimate analysis is useful during mass balance calculations for a thermal or chemical process. Ultimate analysis of wastes is conducted to determine the proportions of carbon, oxygen, hydrogen, nitrogen and sulfur. The determination of the heating value is also important since it indicates the amount of heat obtained when fuel (biomass in the present case) is combusted. Finally, compositional analysis is essential since it determines the conversion efficiency based on the conversion method.

Agricultural biomass can be utilized to produce bio-energy in different ways. At first, it can be used as a directly combustible fuel. Many types of raw or slightly processed biomass produced during agricultural activities are suitable for heat and electricity production. Wood and other plant materials have been directly applied for heating and cooking since the dawn of modern human beings. Before the widespread utilization of fossil fuels, firewood was the main fuel for domestic purposes [22]. Direct combustion of firewood is considered one the most efficient methods to produce bio-energy, although it is not applicable in any heating system because of its bulky size. For these cases, wood chips and wood pellets are great alternatives, demonstrating a volumetric energy content of about 3100 MJ/m^3 and 11,000 MJ/m^3, respectively [22, 23]. Except for wood pellets, the crop residues, grasses and nutshells can also be processed into pellets for their intended usage for bio-energy production. Additionally, the straw, hay, and even cattle dung have been usually collected, dried and burned to prepare warmth and lighting. At the temperature range of 220–300 °C, most of the dry plant materials ignite in the air to cause fire, releasing the inherent bio-energy in heat and light [22]. Finally, charcoal can be produced from wood using pyrolysis to attain a higher energy content material. Good quality charcoal has an energy content of about 60,000 MJ/kg, which is higher than that of coal. It also burns without flame and smoke, giving a temperature as high as 2700 °C [22]. Today, charcoal is still a valuable product, mainly used for heating, cooking, air and water purification, as well as in steel-making. In 2018, over 53 million tons of wood charcoal were produced worldwide, with the leading countries in charcoal production being Brazil, Nigeria and Ethiopia [24]. Production of electricity from agricultural biomass is a highly viable option in developing countries. Usage of biomass to produce electric power is also applied in Europe and North America. The European Union is running various power plants for electricity generation utilizing agricultural biomass [19].

Agricultural biomass can also be used to produce liquid or gaseous biofuels, which in turn can be utilized as a renewable energy source. Depending on the carbon source of biomass feedstock, biofuels can be classified into generations. First-generation biofuels are produced from edible biomass, including starch (from wheat, barley, corn and potato) or sugars (from sugarcane and sugar beet) [25]. Initially, this type of biofuels was considered a promising alternative to offset fossil fuels depletion. However, the usage of edible crops as feedstocks and the impacts on croplands, biodiversity and food supply became a cause of concern soon. For this reason, more sustainable protocols were developed to produce biofuels, which led to the development of the second-generation of biofuels. The combustion of second-generation biofuels is considered to provide neutral or even negative net carbon (emitted-consumed). In this case, the feedstock consists mainly of lignocellulosic materials, which include the abundant and inexpensive nonedible biomass available from plants (agricultural wastes) [25]. Amongst the major agricultural wastes available globally are rice straw, wheat straw, sugarcane bagasse and corn stover (see Table 2.3). From these wastes, Asia generates the highest quantity of wheat and rice straw, while the USA is the leading generator of corn straw and sugarcane bagasse.

Table 2.3 Major agricultural wastes available globally [19]

Agricultural wastes	Quantity (million tons)
Rice straw	731.3
Wheat straw	354.34
Sugarcane bagasse	180.73
Corn stover	128.02

Fig. 2.3 Generations of biofuels [26]. (Reproduced from Ref. [26], https://www.mdpi.com/2071-1050/13/22/12374, under the terms of the CC BY 4.0 license)

The cost-effectiveness of second-generation biofuels still needs development because there are several technical barriers that need to be overcome. Simultaneously, new generations of biofuels appeared (third and fourth generations). More specifically, several kinds of microorganisms started to be used as feedstocks, including macro- and micro-algae, which is related to the development of third-generation biofuels. Furthermore, genetically modified microorganisms such as microalgae, cyanobacteria and fungi are investigated to be utilized as sources for biofuels production, developing the fourth generation of biofuels (see Fig. 2.3).

Among the liquid biofuels produced from agricultural biomass, bioethanol and biodiesel are predominant. Bioethanol can be produced from feedstocks such as wheat, corn and sugar beet through fermentation (first-generation bioethanol) or from lignocellulosic biomass through hydrolysis and subsequent fermentation (second-generation bioethanol). Additionally, it can be produced using thermochemical processes such as gasification followed by either a catalyzed reaction or fermentation [25]. Motor vehicles are the primary application of bioethanol. In conventional combustion engines, it may be used as a transportation fuel either singly or blended with gasoline (commonly blended with gasoline at a low volume percent). Additionally, it can be utilized as a feedstock to produce ethyl tert-butyl ether, which is commonly used as an oxygenated gasoline additive for pollution control. Shifting to the usage of bioethanol can decrease CO_2 concentrations in the atmosphere in two ways. At first, it decreases the dependence on fossil fuels, and second, CO_2 in the atmosphere is consumed during the growth of the feedstock crops [25]. Global commercial bioethanol production accounted for 110 billion liters in 2019, increased more than six-fold from the turn of the century. The five leading countries in bioethanol production are the United States, Brazil, China, India and Canada [27].

Biodiesel, on the other hand, depends mainly on oil crops, and 75% of its production cost is attributed to feedstock production. More than 350 non-edible and edible oil-bearing crops have been proposed as potential feedstock for biodiesel production. Amongst the main ones are the coconut, palm, soybean and sunflower (first-generation feedstocks), and other agricultural remains and wood waste (second-generation feedstocks). Several technologies have been established to produce high-quality biodiesel, including direct usage and blending, pyrolysis of vegetable oil micro-emulsions and transesterification [25]. Today, the global biodiesel market is sized at USD 46,790 million (at the end of 2021), and it is expected to reach USD 51,480 million by the end of 2026, with a CAGR of 5.87% [28].

Agricultural biomass can also be utilized to produce gaseous biofuels. Biogas is a known renewable gaseous fuel alternative to natural gas. It is generated by the anaerobic digestion of organic wastes. Raw biogas consists of 60–65% of methane, 30–35% of CO_2, and small percentages of H_2O vapor, H_2 and H_2S. The upgraded, pipeline-quality biogas (also named biomethane), which is CO_2, H_2S and other contaminants eliminated, is utilized as a natural gas alternative [22]. Only in Asia, more than 3 Tg of methane is used as fuel gas every year [19]. In an effort to bolster the bloc against a looming energy crisis in European Commission countries, the policy has increased the target for domestic biomethane production to 35 billion cubic meters per year by 2030. By 2050, this potential can triple, growing well over 100 billion cubic meters and covering 30–50% of the future European Union (EU) demand for fuel gas [29]. Syngas is another gaseous biofuel produced from the gasification or pyrolysis of plant materials. Chemically syngas consists of 30–60% CO, 25–30% H_2, 5–15% CO_2, 0–5% CH_4, and lower portions of H_2O vapor, H_2S, NH_3 and others, depending on the feedstock type and production conditions. Gasification is commonly used to produce syngas. For example, dry plant biomass is rapidly heated to above 700 °C in a high-temperature combustion chamber of a gasifier and partially burned in the presence of controlled airflow to yield syngas [22]. Today, the major application of syngas is to produce electricity. The global syngas market was sized at $43.6 billion in 2019, and is projected to reach $66.5 billion by 2027, growing at a CAGR of 6.1% from 2020 to 2027 [30].

Table 2.4 provides the present and forecast until 2030 options for biomass utilization for bio-energy production in percentage values. The bio-energy capacity by year in Europe and Worldwide is presented in Fig. 2.4.

Today, a great variety of corporations and organizations around the world are evaluating commercial-scale bio-energy production. Table 2.5 lists ten of the leading bio-energy producers in the world by 2021.

Table 2.4 Bio-energy production from biomass utilization [19]

Biomass utilization for bio-energy production	2006	2010	2020	2030
Heat	94	83.3	71.4	72.3
Biofuels	2.2	10.4	15.7	11.6
Electricity	3.8	6.3	14.9	16.1

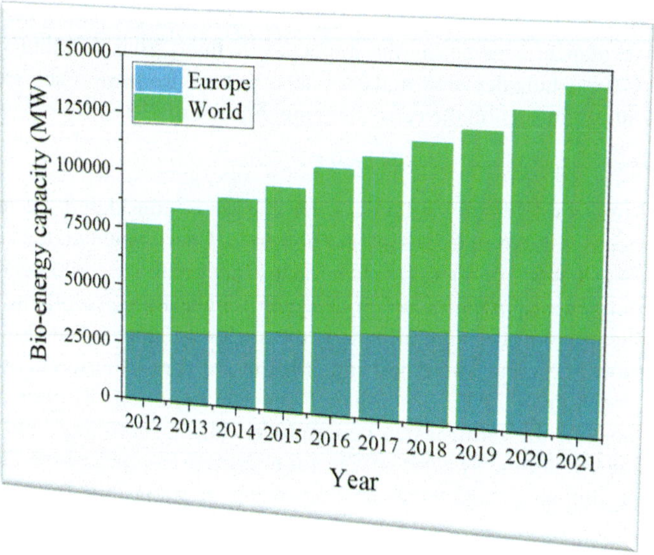

Fig. 2.4 Bio-energy capacity by year in Europe and Worldwide [31]. (The data were obtained freely from the IRENA report entitled "Renewable Energy Statistics 2022", International Renewable Energy Agency, Abu Dhabi)

Table 2.5 List of leading companies (alphabetically short) manufacturing several bio-energy products

Name of the company	Country	Product
Ag Processing Inc	USA	Biodiesel
BlueFire	USA	Bioethanol
Cosan	Brazil	Bioethanol
EnviTec Biogas	Germany	Biomethane
Gevo	USA	Biobutanol
Green Plains	USA	Bioethanol
Montauk Renewables	USA	Biogas
Renewable Energy Group	USA, Netherlands	Biodiesel
Sapphire Energy	USA	Algae biofuels
Verbio	Germany	Biodiesel, bioethanol, biogas

As it is demonstrated, bio-energy products appear as practical, effective and renewable energy resources with continuously increasing economic value, which simultaneously contribute to the reduction of fuel-associated environmental impacts. However, their usage may not always have a net positive environmental and economic attribute. Therefore, their adequate sustainability evaluation for each specific case, incorporating life-cycle thinking, is generally desired. As a part of this approach, life cycle assessment (LCA) can provide integrated results for evaluating their environmental performance, considering the impacts generated from "cradle to grave". In this way, LCA can help decision-makers select the best products or services with the lowest potential impacts on the ecosystem and humanity [32].

Solar Energy

Among all renewable energy sources, solar energy is considered the most viable resource since it is the most abundant and powerful. Every moment, the sun sends to the earth 174 petawatts of energy, which means that one hour could be sufficient to fulfill the energy needs of humanity for one year if it was fully exploitable [33]. Solar energy applications in the agricultural sector are nowadays on the rise, becoming every day more and more affordable and efficient for various farm energy-intensive operations. Solar energy can accomplish the energy requirements of an agricultural system on and off the farm. It can be employed during both day- and night-time (stored in energy storage systems) usage for energy-driven processes, such as crop and grain drying, space and water heating, water pumping, lighting, cooling, etc. [34].

At first, solar energy can be exploited for drying demands in the agricultural sector, significantly reducing drying costs. Conventional drying systems use fuels, such as oil or gas, to produce the appropriate conditions for agricultural products drying. In this case, solar drying can be applied to replace conventional drying systems in multiple modes [35]. In broad terms, solar energy drying systems can be classified into two main groups, namely passive (natural-convection) and active (forced-convection) systems. For both these types of systems, three distinct sub-classes can be identified, depending on the system design and the mode of solar heat utilization. These sub-classes are the integral-type solar dryers, the distributed-type solar dryers and the mixed-mode solar dryers (see Fig. 2.5).

In the case of natural-circulation solar energy dryers (passive systems), their operation depends entirely on solar energy. In these systems, buoyancy forces or wind pressure, acting separately or in combination, circulate air heated by solar irradiation through the crop. Natural-circulation solar-energy dryers appear to be the most attractive option for usage in remote rural locations, and their operation is considered superior and economically competitive to natural open-to-sun drying. Some advantages of these systems compared to the traditional drying techniques are the smaller area of land required to dry similar quantities of crops, higher quantity and quality of dry corps because insects, rodents and fungi are unlikely to infest the crop during drying, while the crop is protected from sudden downpours of rain, shorter drying period compared with open-air drying, thus attaining higher rates of product throughput, combined with relatively low capital and maintenance costs [35]. Considering natural air circulation, at first, there are the integral-type natural-circulation solar-energy dryers (often termed direct solar dryers). In this case, the crop is placed in a drying chamber with transparent walls that allow the insolation necessary for the drying process to be transmitted. The heat produced by solar radiation extracts the moisture from the crop and concomitantly lowers the relative humidity of the resident air, thereby increasing its moisture-carrying capability. Additionally, it expands the

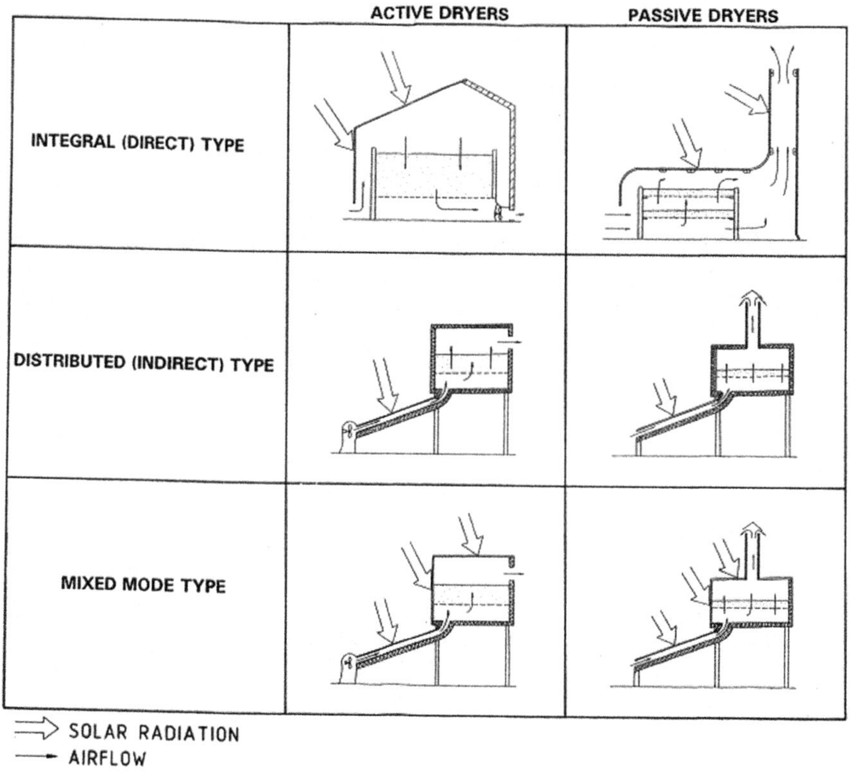

Fig. 2.5 Typical designs of solar energy dryers [35]

air in the chamber, generating its circulation and the subsequent removal of moisture along with the warm air. One of the key benefits of these systems is that no elaborate structures, such as separate air-heating collectors and ducting, are required. However, of the potential drawbacks are the liability of local over-heating, thereby causing crop damage, and the relatively slow overall drying rates achieved due to poor vapor removal [35]. Furthermore, there are the distributed-type natural-circulation solar-energy dryers (often termed indirect passive solar dryers). In these systems, the crop is located in trays inside an opaque drying chamber and heated by warmed air circulation, achieved by a solar thermosyphonic collector. Some advantages of this type of drying are crop protection from direct solar radiation damage and high drying temperatures. On the other hand, some shortcomings include the temperature fluctuations in the air leaving the air heaters, thereby making it difficult to maintain constant operating conditions within the drying chamber, and operational difficulties such as the difficulty of loading and unloading the trays and occasional stirring of the drying product. Besides, they are also more complex and expensive systems compared to integral-type natural-circulation solar energy dryers [35]. Finally, there are mixed-mode natural-circulation solar-energy dryers. These systems combine the features of the integral (direct) type and the distributed (indirect) type natural-circulation solar-energy dryers. In this case, the required heat for the drying process is provided by the combined action of sunlight radiation that is incident directly on the product as well as by a solar air heater [35].

On the other hand, active solar drying systems depend only partly on solar energy. These systems employ solar energy and/or fossil-fuel-based or electrical heating and motorized fans and/or pumps for air circulation. Thus, active solar dryers are considered forced-convection dryers [35]. For active solar drying, at first, there are the integral-type active solar energy drying systems. These are solar drying designs in which the solar energy collection unit is an integral part of the entire system. Thus, there is no requirement for special ducting to channel the drying air to a separate drying chamber. Three distinct designs of integral-type active solar dryers can be identified. These include the direct absorption dryers, where the product absorbs solar radiation directly, thus no separate solar collector is required; the solar collector-roof/collector-wall dryers that are usually storage-type dryers, where, in these designs, the solar collector forms an integral part of the roof and/or wall of the drying/storage chamber; and the internal-absorber-chamber greenhouse dryers, which usually consist of a transparent exterior that acts as the solar collector glazing and an inner drying chamber that is also the absorber [35]. Furthermore, there are distributed-type active solar-energy drying systems. In these systems, the solar collector and drying chamber are separate units. In this case, the efficiency of solar collectors decreases with the increase of outlet temperatures. Thus, a critical decision in the design of distributed active solar dryers would be either to choose high drying air temperatures applying low air-flow rates or low-temperature drying, thus minimizing the cost of insulation since heat losses are much lower. Most distributed-type active solar drying systems have similar structural designs comprising the basic components, although modifications to the typical design have tended to be based on the solar air heaters, air recirculation and fan/pump location [35]. Finally, there are the mixed-mode active solar-energy dryers, which combine features of the integral and distributed types, although these are rather uncommon designs. To achieve more efficient energy usage, some active solar dryers are equipped with thermal storage devices. This improves drying during periods of low insolation levels or night-time. Desiccants are incorporated in some designs to further reduce the relative humidity of the drying air to improve its moisture-carrying capacity. The use of desiccants would only be appropriate for forced-convection systems, as their incorporation into the system increases the resistance to airflow. Finally, some active solar dryers employ air-heating solar collectors as supplements to electricity or dehydrators operated using fossil fuels to reduce the overall conventional energy consumption (often termed "hybrid systems") [35].

Solar energy can also be used to accomplish the electric power requirements in agricultural systems through the employment of photovoltaic technologies, while the excess of the produced energy can be sold to the grid, providing an extra income to the farmers. In essence, the aforementioned approach regards the co-usage of land for both agriculture and photovoltaic systems installation, often called agrivoltaics. The concept of agrivoltaics dates back to the 1980s decade [36]. This approach is considered increasingly popular in our day as an emerging and very promising strategy to tackle the challenges facing humanity today for increased energy and food requirements. This deliberate co-location of agriculture and photovoltaic systems is intended to alleviate land usage competition and boost revenues for landowners, among other benefits. In this way, on- and off-farm activities can be powered through the electricity provided by solar cells, contributing simultaneously to the enhancement of the rural electrification rate, which is considered quite important in our days for increasing the living quality of the rural population [13]. Numerous empirical and practical studies have already investigated the technical viability of agrivoltaic systems, examining the simultaneous light-to-electricity conversion and plant cultivation or livestock production, and the corresponding effect of one on the other. Depending on the facility scale and climatic conditions, the energy produced by photovoltaics can be used for various tasks in the agricultural sector, including heating, cooling and lighting. Under the agrivoltaics approach, photovoltaic systems are usually mounted on agricultural lands, above the ground to facilitate cultivation practices or on agricultural structures such as greenhouses. Another possibility is to power farm machinery or vehicles, although a key challenge, in this case, is the low capacity of the energy storage systems applied, making long-time operation of electric-powered machinery difficult with a single charge [34]. Overall, it has been demonstrated that agrivoltaic systems are a technically and economically viable way to utilize agricultural land, capable of overcoming the dominant separation of food and energy production, while also even boosting land productivity [37].

Solar energy can also be used for irrigation demands in agriculture. The agricultural sector is the major water consumer in the world and accounts for approximately 70% of freshwater consumption. From this number, most crop production still comes from rainfed agriculture, where crop requirements are fulfilled from the rainfall water held in the root zone of the soil [34]. On the other hand, water pumping systems are usually employed to provide additional water volume when the rainfall water is not adequate to ensure the desired crop quantity and quality. For this case, the conventional pumping systems applied usually operate using power produced from fossil fuels. As an efficient and sustainable alternative, solar-powered pumping systems are nowadays increasingly applied for watering crops, minimizing the dependence on diesel, gas or coal electricity, ultimately providing an environmentally sustainable solution to address climate change. Typical solar-powered pumping systems consist of a photovoltaic array to convert solar energy into electricity, a motor that converts the electricity to mechanical energy, and a water pump, where the mechanical energy is converted to hydraulic [38]. In most cases, an energy storage system is also employed to store the dc electric power for powering demands when the sunlight irradiation is not adequate (e.g., in cloudy conditions, night-time, etc.) or for further uses.

Solar-powered water pumping systems can be classified into three main types according to their application, namely submersible pumps, surface pumps and floating water pumps. The submersible pumps are used to draw water from deep wells, the surface pumps are used in the case of water drawn from shallow wells, springs, ponds, rivers or tanks, and the floating water pumps are used to draw water from reservoirs with adjusting height ability [38]. Additionally, solar-power water pumps may be classified under two categories depending on the operating principle of the pump. These are the dynamic pumps and the positive displacement pumps. Dynamic pumps operate by developing a high liquid velocity and pressure in a diffusing flow passage. On the other hand, positive-displacement pumps operate by forcing a set amount of fluid from the inlet pressure section of the pump into its discharge zone [38]. In general, the efficiency of dynamic pumps is considered lower as compared to positive displacement pumps, but the former ones have comparatively lower maintenance requirements. Some advantages of solar-powered pumping systems are the ability to offer water pumping in remote locations where electric grids are not available, as well as the lower cost compared to diesel-powered pumps. On the other hand, the main disadvantages of this technology are that they are only suitable for low-lift small-scale irrigation, while the energy supply in cloudy seasons and night-time is minimal or absent. The performance of solar-powered water pumps depends on several parameters. These include the solar radiation availability at the location, the total dynamic head (total equivalent height that fluid is to be pumped, friction losses in the pipe), the flow rate of water, the total quantity of water requirement and the potential energy required in raising the water to discharge level (hydraulic energy) [38].

Solar-powered water pumping systems have shown significant advancements in the last few years. The first-generation solar-powered pumping systems used centrifugal pumps, usually driven by dc motors and variable frequency ac motors, which demonstrate long-term reliability and hydraulic efficiency varying from 25 to 35%. The second-generation pumping systems use positive displacement pumps, diaphragm pumps or progressing cavity pumps, generally characterized by low input power requirements, low capital cost and high hydraulic efficiencies of even 70%. The current most advanced solar pumping technology uses electronic systems that further increase the output power and overall efficiency of the system. Combined with the advancements in the tracking photovoltaic systems, the pumping systems are now able to extend the time for peak water yield. Today, the solar-powered water pumps available in the market can lift water from 5 m to higher than 200 m, with up to 250 m^3/day outputs [38].

Besides the aforementioned, the usage of solar energy for the production of fertilizers is another emerging approach today. The conversion of triple-bonded molecular N_2 from the air into fixed nitrogen products that act as nutrients using solar energy presents a great opportunity to develop "solar fertilizers" [39]. This strategy shares several similarities with the development of solar fuels but also has some unique advantages and challenges. Today, a vast industry exists to produce and deliver nitrogen-based fertilizers to farms, which benefit from higher crop yields. Yet, this undermines the sustainable development of the agricultural sector since, in many cases, excess chemical runoff often spills into rivers and coastal waterways. A promising strategy for overcoming the drawbacks of the use of traditional fertilizer is by utilizing solar energy to decentralize production. Solar fertilizers are an alternative for local fertilizer production by utilizing solar energy, nitrogen, water and oxygen from the ambient air to produce low-concentration ammonia- or nitrate-based fertilizers at or near farms where they are going to be used. This is advantageous because it gives way for the direct capture of solar energy in a chemical product that can be utilized close to the place of production, avoiding long-distance transportation or storage of electrical energy in batteries [39].

Wind Energy

Wind energy is one of the fastest-growing renewable energy sources in the world. From many, it is regarded as the most affordable and clean source of energy and is anticipated to play an increasingly important role in the future worldwide energy scene. Actually, the wind is a form of solar energy. The heating of the atmosphere by the sun, the earth rotation and irregularities, all contribute to the development of winds. For as long as the sun shines and the wind blows, wind power can be used to be converted into useful energy. Among the main advantages of wind energy is primarily the low wind power cost. The land-based utility-scale wind is one of the lowest-priced energy sources available today. The global weighted-average cost of electricity in 2019 was USD 0.053/kWh for new onshore wind farms, with country/region values of between USD 0.051 and USD 0.099/kWh. Without funding, costs for the most competitive projects are currently as low as USD 0.030/kWh [40].

Wind energy and agriculture pair well. For centuries, wind power was used in the agricultural sector for grind processing and pumping. However, as electricity and fossil fuels became cheaper, harnessing wind power for these purposes became less popular. Today, as the sustainability challenges become more and more apparent, the strategy of integration of wind power systems in agriculture comes back, in line with the worldwide effort taken to support renewable energy. Under the rapid technological advancements, farmers now have the opportunity to raise their crops and livestock, while simultaneously use their land to develop wind farms that can be employed for either meeting the energy demands of their agricultural systems or to provide them with an extra income by selling the produced electric power to the grid. The latter is considered very important, especially for small farmers, who have to deal with the increasing prevalence of big agribusiness and cheap imported food in order to survive. Amongst the main advantages of wind energy for agriculture is that wind turbines can be installed in the middle of agricultural land with minimal crop damage. On the other hand, the main disadvantage is that the wind is a fluctuating source of energy, a characteristic that greatly affects the power output. Besides, the unwanted noise that is produced by the wind turbines poses a threat to wildlife, causing even a visual disturbance [13].

One of the earliest applications of wind energy in agriculture is related to griding processing. Since ancient times, wind turbines (called windmills) have been used as a primary method to grind grains into flour by using wind power. Today, wind power is employed for grain and wheat grinding, mainly in developing countries. The main advantages of griding using wind energy are that it is easy to operate, there are no fuel costs, and it is an approach free from creating greenhouse gases. Amongst the main disadvantages is that it is considered an inefficient technique for grinding grains compared to other technology gridding systems, while it also creates noise pollution [13].

Among the most famous applications of wind energy in the agricultural sector is also for irrigation purposes and water supply (see Fig. 2.6). Wind pumps frequently provide a cost-effective solution for domestic and community water supply, small-scale irrigation, and livestock water usage in off-grid places when there is adequate wind (>5 m/s) and groundwater supplies. In order to select a suitable windpump, the daily water requirement, mean wind speed, total pumping head, well drawdown, water quality and storage requirements need to be considered [41]. According to the end-use, different windpumps should be developed and applied because of the different technical, operational and economic requirements. In general, windpumps applied for water supply have to be ultra-reliable to run unattended for most of the time. They should also provide minimum requirements for maintenance, while they should be capable of water pumping from depths of 10 m and more. A typical windpump for farming should be able to run for over 20 years, with its maintenance to be applied no more than once per year, without any major replacements. This is four to ten times the lifetime of most small diesel engines or about 20 times the lifetime of a small engine pump. Therefore, windpumps of this standard are usually fabricated with steel components and drive piston pumps via reciprocating pump rods. For this reason, these systems are usually considered expensive in relation to their power output [42]. On the other hand, windpumps developed for irrigation are usually of lower cost since they do not have such demanding technical requirements. Irrigation duties are seasonal and windpumps may only be used for a limited fraction of the

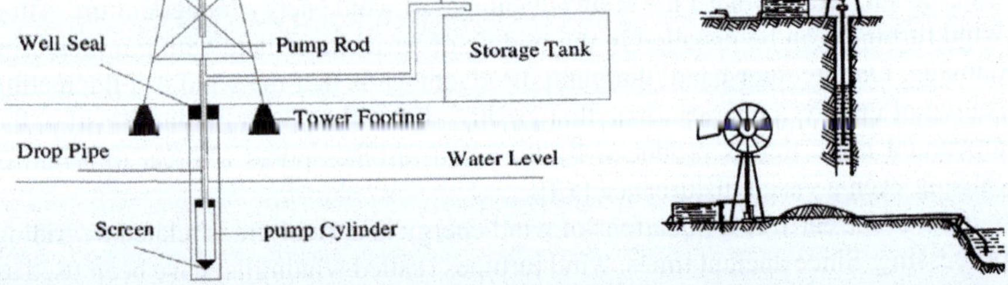

Fig. 2.6 **a** Components of a windpump, **b** Typical farm windpump installation configurations, including borehole to raised storage tank, well to surface storage tank and surface suction pump [42, 43]. (Figure 2.6a is reproduced from Ref. [43], https://journals.sagepub.com/doi/full/10.1177/1687814019880405, under the terms of the CC BY 4.0 license, Fig. 2.6b is reproduced with permission from FAO)

year. Additionally, irrigation usually involves the farmer being present. Thus, it is not so critical to have a machine capable of running unattended. Therefore, windpumps used for irrigation in the past tend to be indigenous designs that are often constructed or improvised by the farmer as a kind of low-cost mechanization [42].

Table 2.6 lists the factors affecting the efficiency of a wind-pump. Depending on the efficiency factor, the conversion of the wind energy (using the mean wind speed) to hydraulic energy is determined between 7 and 27%. Because the wind speed continuously varies, the actual energy in the wind is considerably greater than if the wind blew continuously at the mean speed. The exact proportion by which the energy available exceeds the figure that can be calculated based on the mean speed depends on the shape of the wind velocity distribution curve. Yet, it is typically twice the figure arrived at using the mean speed. This is usually offset by applying an efficiency factor of about 200% to avoid the underestimation of the total efficiency of the energy conversion.

Finally, wind power can be employed to produce electric power. Electricity produced by wind turbines can be used to cover the energy demands in agricultural systems, arising mainly from heating, cooling, lighting, etc. [13].

Hydropower

Hydroelectric power is of the most mature, reliable and cost-effective renewable energy sources available today. In the last decade, hydropower accounted, on average, for the production of around 16% of the world's electricity and more than 4/5 of the world's renewable energy supply [44]. Hydropower facilities are usually classified into three types, namely impoundment, diversion and pumped storage. The most prevalent kind of hydroelectric power facility is an impoundment. In this case, a dam is used to store river water in a reservoir. Released water from the reservoir spins a turbine as it passes through, activating a generator to produce electricity. On the other hand, a diversion, which is sometimes also called "run-of-river" facility, channels a segment of a river through a penstock and/or a canal in order to utilize the natural decline of the riverbed elevation to produce energy. A penstock is a sealed conduit that channels the water flow to turbines, where water flow is regulated by gates, valves and turbines. This type of facility may not require the usage of a dam. Finally, pumped storage hydropower facilities work like giant batteries. When the electricity demand is low, they are able to store energy by pumping water from a lower reservoir

Table 2.6 Factors affecting a windpump system efficiency [42]

Factor	Typical efficiency
Rotor to shaft	25–30
Shaft to pump	92–97
Pump	60–75
Wind power capture	30–50
Actual wind energy to wind energy calculated using mean wind speed	180–250
Total energy conversion efficiency based on mean wind speed	7–27%

to an upper reservoir, while, during periods of high demand for electricity, the water is sent back to the lower reservoir and rotates a turbine, generating electricity [45]. Amongst the main advantages of the usage of hydropower are the low operating and maintenance costs of hydro-powered water pumps, as well as the reliability of these systems since they are composed of few moving parts. However, despite being a sound alternative to conventional water pumping systems, they are not popular due to the limited number of locations that can be developed and the lack of suitable water flows. Besides, the energy supply through hydro-powered water pumps is of high variation and disturbance due to seasonal variations in streamflow [13].

Hydropower can be employed in the agricultural sector to provide useful energy for the operation of agricultural systems, contributing simultaneously to secure independence from fossil fuels. Small-scale hydropower, also referred to as micro-hydro, can be an important source of energy to supplement a farm, harnessing the energy of flowing water and converting that energy to mechanical or electrical use. Typical micro-hydro installations are designed to produce up to 100 kW of electricity using the natural flow of water. On the other hand, installations below 5 kW, also called pico-hydro, can be employed to provide electricity for agro-processing facilities in remote communities, where neither connection to the grid nor micro-hydropower is feasible. The categories of hydro-powered plants and their usual applicability are tabulated in Table 2.7.

In 2020, the global hydropower market was sized at US $214.34 billion [48]. The global installed hydropower capacity increased from 1010 GW in 2010 to 1330 GW in 2020, setting a CAGR of 2.7% for the aforementioned decade, while it is anticipated to reach approximately 1600 GW by 2030 [49]. The levelized cost of electricity (LCOE) of hydropower remained stable throughout the past decade, at about USD 0.04/kWh, setting it among the lowest-cost sources of electricity (see Fig. 2.7) [50].

Geothermal Energy

Geothermal energy, namely the flow of heat energy radiating from the earth's core, provides unique opportunities for cost-efficient, sustainable food production and processing [51]. Geothermal power is considered a sustainable and renewable energy source because the heat extraction is small compared with the earth's heat content,

Table 2.7 Categories of hydro-powered plants and their corresponding applicability [46]

Type	Power output	Applicability
Large	>100 MW	Large urban population centers
Medium	10–100 MW	Medium urban population centers
Small	1–10 MW	Small communities with the possibility to supply electricity to a regional grid
Mini	100 kW–1 MW	Small factories or isolated communities
Micro	5–100 kW	Small isolated communities
Pico	<5 kW	1–2 houses

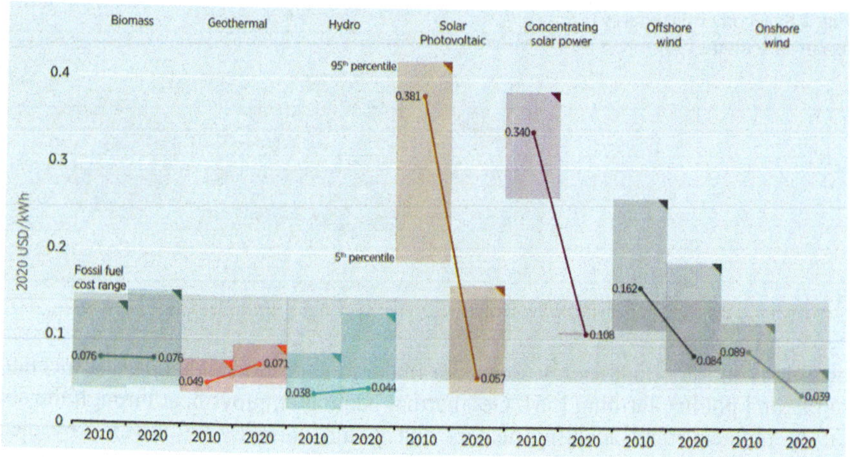

Fig. 2.7 Global levelized cost of electricity values from commissioned, utility-scale renewable power generation technologies, 2010–2020 [47]

while the greenhouse gas emissions of conventional geothermal electric stations are, on average, less than 5% of that of conventional coal-fired plants. Amongst the advantages of geothermal energy is that it can be used directly (such as for heating, cooking, etc.), its usage does not depend on weather conditions, and there is no wastage or generation of by-products. Besides, the maintenance cost of geothermal power plants is much less compared to fossil-fired power plants of the same capacity [52]. According to estimates, the geothermal energy market is expected to grow from $55.80 billion in 2021 to $83.27 billion in 2028, at a 5.9% CAGR during 2021–2028 [53].

Regarding the agricultural sector, geothermal energy is starting to become more and more prevalent. The applications of geothermal energy in agriculture are similar to those of solar energy; however, the former is available 24 h per day and 365 days per year, making it more convenient to access, especially at higher latitudes that receive less daylight. Geothermal energy can be employed for many purposes in agriculture, such as heating and cooling buildings, soil and water, drying crops and grains, as well as for heating greenhouses (see Fig. 2.8).

Greenhouse heating using geothermal energy is one of its most common direct applications that has been extensively investigated during the last 25 years [52]. It can be employed to grow crops such as vegetables and fruit in addition to flowers, houseplants and tree seedlings. Moreover, geothermal district heating and cooling can be employed to provide heat to buildings and industries through a distribution network. Geothermal energy can be employed for space heating and cooling, industrial process heat requirements and domestic hot water supply by district heating using peaking stations, distribution systems, central pumping stations and in-building equipment. Replacing traditional heating methods in agricultural systems with geothermal heating can sustainably reduce the costs for the producer since space

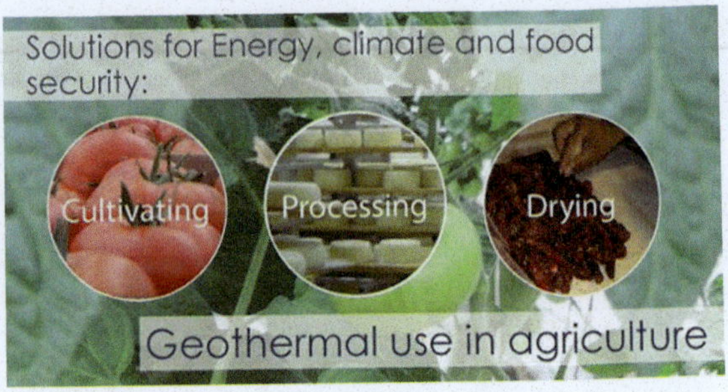

Fig. 2.8 Geothermal energy use in agriculture [54]

heating is an important energy consumer in the food production industry, especially in pig and poultry farming [55]. Geothermal heating employs heat through the soil or air, replacing and adapting heaters that would normally be powered by electricity or fossil fuels. Besides, geothermal ventilation systems can be applied in replacement of the conventional non-renewable-powered systems that are used to improve air quality for promoting optimal health and growth of animals. Employing non-renewable resources, such as fossil fuels, to replace contaminated indoor air can require a tremendous amount of energy, most of which can be provided by geothermal-powered air/space dehumidifiers and ventilation systems [56].

The drying of crops and animal products is also one of the most versatile and simplistic applications of geothermal energy currently being practiced in the agricultural and food industry. These drying procedures can range from "do it yourself" methods to complex industrial-sized geothermal-powered dehydrators. Herein, geothermal energy excels from solar drying, especially in areas that receive low solar radiation, such as at higher elevations and in colder climates. Many countries currently use geothermal energy on an industrial scale for drying various grains, vegetables and fruit crops, such as wheat, cereals and onions. Today, the main consumers of this renewable resource for drying purposes are China, the USA and Hungary [55].

Dairy processing is yet another application of geothermal energy. Geothermal steam can be employed in dairy factories, replacing unsustainable processes, such as coal burning, for milk powder production, while fresh milk from the farmers can be pasteurized using hot geothermal water or geothermal brine as heating liquids [52].

Geothermal water can also be used in open-field agriculture to keep the soil at a steady warm temperature. In conventional agriculture, farmers install irrigation pipes under the soil to provide both water and heat to the crops. Herein, the usage of geothermal water can extend the growing season and protect the plants from being damaged by low air temperatures. Geothermal water can also be used to sterilize soil to kill pests, fungi and diseases that can damage crops. In this case, the farmers either heat the soil from pipes underneath it or apply the steam above the soil, covering it with a plastic sheet to retain the heat inside [52].

Finally, the usage of geothermal systems can even be considered a sustainable solution for freshwater production. Geothermal water desalination is a simple method to secure high-quality freshwater that can be used for a variety of applications [55].

Challenges to the Widespread Adoption of Renewable-Energy-Powered Agriculture

Adopting renewable energy systems in the agricultural sector is considered an emerging and valuable strategy recruited today to deal with the ever-increasing food and energy demands. However, despite the arising multifaceted benefits that can lead to a significant step closer to the sustainable development of humankind on this planet, their widespread adoption has not occurred yet, and the level at which this development agenda is prioritized varies greatly from country to country. This arises from a number of barriers that need to be overcome, mainly related to economic, technical, informational and political challenges (see Fig. 2.9).

Economic challenges are considered the main barriers to the universal adoption of renewable-energy-powered agriculture. This mainly lies in the high initial cost of "converting" or "changing practices", including the cost of new equipment and the expenses of switching from one management style to another. These costs, combined with the declined net farm income and the relatively long payback time for many types of renewable energy systems, result in increased risks and the uncertainty of profitability that farmers cannot afford without substantial government support or additional private investments. On the other hand, government financial programs are often ineffective since they are tailored for large farms or show bias toward large-scale agricultural production. Low-income farmers are typically unable to invest in large-scale renewable-energy-powered agricultural systems, while a renewable-energy-based agricultural development is considered financially unsustainable for many local institutions in the long financial run, especially in developing countries. Besides, under the policy-stimulated incentives, long-term commitments seem uncertain and

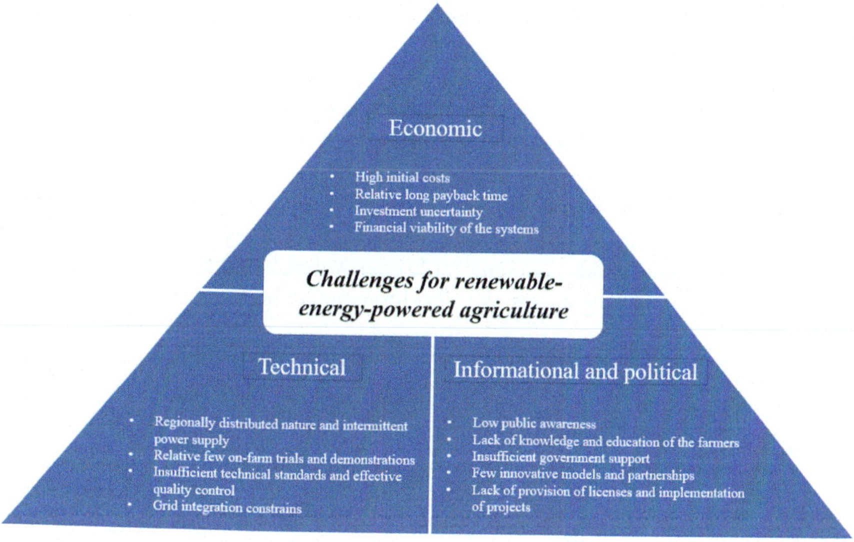

Fig. 2.9 Challenges to the widespread adoption of renewable-energy-powered systems in the agricultural sector [13, 14]

the net profit unpredictable. The competitive cost-effectiveness of renewables is also disputed for those requiring significant infrastructure development and technically skilled laborers for ongoing maintenance.

With the recent developments and vast progress in renewable energy technologies, the application of renewable energy systems in agriculture has become more feasible. However, technical challenges still exist that slow down their widespread adoption in the agricultural sector. The shortage of funding for renewable-energy-powered agriculture discourages investors from undertaking crucial first steps, such as energy resource assessments or feasibility studies for renewable energy projects for agriculture. In many cases, on-farm trials and demonstrations lack technical standards and effective quality control for the simultaneous agricultural products development and energy conversion and exploitation are absent, while there is also a shortage of qualified personnel, such as policy analysts, economic managers and engineers, to undertake these projects. All these act as deterrents for the universal adoption of renewable energy solutions in agriculture. Besides, the infrastructure to support renewable energy systems is often inadequate. Grid integration of off-grid networks powered by renewable energy sources is a challenge in many countries, especially in developing ones. Furthermore, in many agricultural renewable-powered systems, the output energy stability and reliability are considered insufficient compared to conventional systems.

The challenges are not only economic and technical but also informational and political. Public awareness of the environmental and economic benefits of renewable energy sources adoption in the agricultural sector is relatively low, especially in developing countries, which delays the formation of a comprehensive energy policy. Many farmers lack knowledge and education concerning sustainable agricultural development, and they express uncertainty about changing to something unknown, while some of them believe that sustainable practices lead to reduced yields. The public is still unable to motivate the local and national governments to take pivotal initiatives for the development of agricultural renewable energy systems and technologies. Excluding a few developed countries, the present government policies do not reasonably support renewable energy by providing sufficient budgetary allocations and/or a legislative framework that can attract private or foreign investments, and they continue to resist feed-in tariffs. Besides, the lack of provision of licenses and implementation of projects in renewable energy development leads to the unsuccessful adoption of strategies for the development of renewable-energy-powered agriculture.

2.4 Outlook

The energy demands of the agri-food system today are estimated at about one-third of the total global energy production and are forecast to greatly increase in the next years, owing to the rapid population growth. Fossil fuels are considered the unsung hero of the agricultural revolution and, until now, remain the dominant energy lifeblood of the agricultural sector. However, climate change and the rapid depletion of fossil fuels are among the biggest challenges that humanity faces today, and now, more than ever, there is an imperative need to adopt anticipatory strategies to ensure WEFE nexus sustainability. To this aim, the exploitation of renewable energy technologies in farming is considered a win–win-win solution. Together with energy-efficient technologies, the exploitation of biomass, solar, wind, hydropower and geothermal energy to power a large variety of on-farm and off-farm activities can be employed to enhance green energy production and help toward the mitigation of food scarcity and climate change. Yet, still today, the adoption of sustainable agriculture practices and technologies is limited due to several barriers derived mainly from economic, technical, informational and political challenges. Now more than ever, it is urgent for humanity to raise awareness of the impending nexus crisis, overcome the arising challenges and establish the mentality of sustainability. Herein, the role of the scientific community is significant, amongst others, to propose innovative and effective solutions for the development of sustainable, green-powered agriculture.

References

1. Brodt S, Six J, Feenstra G et al (2011) Sustainable agriculture. Nat Educ Knowl 3:1
2. Sivakumar B (2021) Water-energy-food nexus: challenges and opportunities. Stoch Environ Res Risk Assess 35:1–2
3. Bundschuh J, Chen G (2014) Sustainable energy solutions in agriculture, first. CRC Press Taylor & Francis Group
4. Betts N (2015) Introduction to sustainable agriculture
5. Wikipedia (2022) Triple bottom line. https://en.wikipedia.org/wiki/Triple_bottom_line. Accessed 28 Jun 2022
6. Food and Agricultural Organization of the United Nations (FAO) (2011) Energy-smart food for people and climate. Rome
7. IRENA and FAO (2021) Renewable energy for agri-food systems: towards the sustainable development goals and the Paris agreement. Abu Dhabi and Rome
8. Gro Intelligence (2019) A Gro use case: modeling energy needs in agriculture
9. Eurostat (2019) Agri-environmental indicator - energy use
10. Noronha T (2019) Alternative livelihoods working glossary
11. Schnepf R (2004) Energy use in agriculture: background and issues
12. Carbon Trust (2019) Energy efficiency in agriculture
13. Rahman MM, Khan I, Field DL et al (2022) Powering agriculture: present status, future potential, and challenges of renewable energy applications. Renew Energy 188:731–749
14. Rodriguez JM, Molnar JJ, Fazio RA et al (2008) Barriers to adoption of sustainable agriculture practices: change agent perspectives. Renew Agric Food Syst 24:60–71
15. Alhazmi H, Loy ACM (2021) A review on environmental assessment of conversion of agriculture waste to bio-energy via different thermochemical routes: current and future trends. Bioresour Technol Rep 14:100682
16. Fortune Business Insights (2020) Market research report (bioenergy market)
17. European Biomass Industry Association (2021) Bioenergy
18. Garba A (2020) Biomass conversion technologies for bioenergy generation: an introduction. In: Biotechnological applications of biomass. IntechOpen
19. Saleem M (2022) Possibility of utilizing agriculture biomass as a renewable and sustainable future energy source. Heliyon 8:e08905

20. Lal R (2005) World crop residues production and implications of its use as a biofuel. Environ Int 31:575–584
21. Gupta GK, Mondal MK (2020) Bioenergy generation from agricultural wastes and enrichment of end products. Refin Biomass Residues Sustain Energy Bioprod Technol Adv Life Cycle Assessment, Econ 337–356
22. Guo M, Song W, Buhain J (2015) Bioenergy and biofuels: history, status, and perspective. Renew Sustain Energy Rev 42:712–725
23. Forest Research Typical calorific values of fuels. https://www.forestresearch.gov.uk/tools-and-resources/fthr/biomass-energy-resources/reference-biomass/facts-figures/typical-calorific-values-of-fuels/. Accessed 29 Jun 2022
24. Food and Agricultural Organization of the United Nations (FAO) (2021) Forest product statistics
25. Alalwan HA, Alminshid AH, Aljaafari HAS (2019) Promising evolution of biofuel generations. Subject review. Renew Energy Focus 28:127–139
26. Khan N, Sudhakar K, Mamat R (2021) Role of biofuels in energy transition, green economy and carbon neutrality. Sustainability 13:12374
27. Hoang T-D, Nghiem N (2021) Recent developments and current status of commercial production of fuel ethanol. Fermentation 7:314
28. Market Data Forecast (2022) Biodiesel market
29. European Biogas Association (EBA) (2022) The biomethane sector will deliver 20% of current EU gas imports from Russia by 2030. https://www.europeanbiogas.eu/the-biomethane-sector-will-deliver-20-of-current-eu-gas-imports-from-russia-by-2030/. Accessed 1 Jul 2022
30. Allied Market Research (2021) Syngas market 2021
31. IRENA (2022) Renewable energy statistics 2022
32. Prasad S, Singh A, Korres NE et al (2020) Sustainable utilization of crop residues for energy generation: a life cycle assessment (LCA) perspective. Bioresour Technol 303:122964
33. Kabir E, Kumar P, Kumar S et al (2018) Solar energy: potential and future prospects. Renew Sustain Energy Rev 82:894–900
34. Tariq GH, Ashraf M, Hasnain US (2021) Solar Technology in Agriculture. In: Ahmad F, Sultan M (eds) Technology in agriculture, first. IntechOpen, London
35. Ekechukwu OV, Norton B (1999) Review of solar-energy drying systems II: an overview of solar drying technology. Energy Convers Manag 40:615–655
36. Goetzberger A, Zastrow A (1982) On the coexistence of solar-energy conversion and plant cultivation. Int J Sol Energy 1:55–69
37. Dupraz C, Marrou H, Talbot G et al (2011) Combining solar photovoltaic panels and food crops for optimising land use: towards new agrivoltaic schemes. Renew Energy 36:2725–2732
38. Chandel SS, Nagaraju Naik M, Chandel R (2015) Review of solar photovoltaic water pumping system technology for irrigation and community drinking water supplies. Renew Sustain Energy Rev 49:1084–1099
39. Comer BM, Fuentes P, Dimkpa CO et al (2019) Prospects and challenges for solar fertilizers. Joule 3:1578–1605
40. IRENA Wind Power. https://www.irena.org/costs/Power-Generation-Costs/Wind-Power. Accessed 4 Jul 2022
41. Energypedia (2018) Wind energy—introduction. https://energypedia.info/wiki/Wind_Energy_-_Introduction. Accessed 5 Jul 2022
42. Food and Agriculture Organization of the United Nations (FAO) (1986) Water lifting devices. Rome
43. Aized T, Sohail Rehman SM, Kamran S et al (2019) Design and analysis of wind pump for wind conditions in Pakistan. Adv Mech Eng 11:1–18
44. International Renewable Energy Agency (IRENA) (2012) Hydropower
45. Energy Efficiency & Renewable Energy Types of Hydropower Plants. https://www.energy.gov/eere/water/types-hydropower-plants. Accessed 5 Jul 2022
46. Sustainable Sanitation and Water Management Toolbox (2020) Hydropower (Small-scale). https://sswm.info/water-nutrient-cycle/water-distribution/hardwares/water-network-distribution/hydropower-%28small-scale%29. Accessed 6 Jul 2022

47. International Renewable Energy Agency (IRENA) (2021) Renewable power generation costs in 2020
48. Maximize Market Research (2021) Hydropower market- industry analysis and forecast (2019–2027): by application, and by region
49. GlobalData (2022) Power generation and cumulative capacity of hydro power plants in China (2017–2021). https://www.globaldata.com/data-insights/hydro-power-generation/power-generation-and-cumulative-capacity-of-hydro-power-plants-in-china-2017-2021/. Accessed 5 Jul 2022
50. Statista (2022) Levelized cost of electricity (LCOE) of hydropower worldwide from 2010 to 2020. https://www.statista.com/statistics/799349/lcoe-of-hydropower-worldwide/. Accessed 5 Jul 2022
51. Food and Agriculture Organization of the United Nations (FAO) (2015) For developing countries, new opportunities in geothermal energy. https://www.fao.org/news/story/pt/item/281565/icode/. Accessed 6 Jul 2022
52. Omenikolo AI, Obasi I (2020) Application of geothermal energy in agriculture: a review. Int J Eng Appl Sci Technol 5:666–673
53. Fortune Business Insights (2022) Geothermal energy market size, share & COVID-19 impact analysis by type, by application, and regional forecast, 2021–2028
54. Think Geoenergy (2022) EGEC: geothermal energy use in agriculture. https://www.thinkgeoenergy.com/egec-geothermal-energy-use-in-agriculture/. Accessed 6 Jul 2022
55. Thermtest Instruments (2021) Applications of geothermal energy in the agriculture industry and the influence of material thermal conductivity. https://thermtest.com/applications-of-geothermal-energy-in-the-agriculture-industry-and-the-influence-of-material-thermal-conductivity. Accessed 6 Jul 2022
56. Bundschuh J, Chen G, Tomaszewska B, et al (2017) Solar, wind and geothermal energy applications in agriculture: back to the future? In: Geothermal, wind and solar energy applications in agriculture and aquaculture, first. CRC Press

Chapter 3
Solar Photovoltaic Energy in Agriculture

"Solar photovoltaic energy in agriculture" is the main thematic content accounted for in the present book and the main topic for discussion in this chapter. For readers' benefit and a comprehensive presentation of the current state of the art on the agrivoltaics approach, seven levels for discussion are considered. Initially, an introduction to agrivoltaics takes place, where the reader acquires the minimum background information for the topic, while afterward, the application modes of photovoltaics in agriculture and results from case studies on the corresponding impact of their application on crop yields and quality are presented. Subsequently, a rough overview of the costs for photovoltaics-farming dual land-use compared to bare ground-mounted PV systems is given, followed by a brief report on up-to-date information on the emerging market of agrivoltaics. Finally, the significance of photovoltaic agriculture in the Mediterranean area for sustainable development is demonstrated, while the barriers and opportunities for further development of agrivoltaics are also discussed.

3.1 Introduction to Agrivoltaics

Renewable energy is rapidly advancing in our days as a countermeasure to control climate change and the great rising global energy crisis. The energy transition from fossil fuels to renewable energy is considered today one of the most urgent and effective actions that should be taken toward sustainability on our planet. Amongst the alternatives, photovoltaic (PV) energy is considered a key part of the solution as it was also recognized by the delegates at the 26th United Nations Climate Change Conference of the Parties (COP26) that took place in Glasgow, Scotland, United Kingdom in 2021, where solar electricity production was mentioned to play a leading role in the energy transition process [1].

At this conference, the COP26 summit brought parties together (around 30,000 delegates from governments, institutions and technology companies) to accelerate

D. A. Chalkias and E. Stathatos, *The Emergence of Agrivoltaics*,
Green Energy and Technology, https://doi.org/10.1007/978-3-031-48861-0_3

action toward the goals of the Paris Agreement and the United Nations Framework Convention on Climate Change (Sustainable Development Goal 7 (SDG7)). In this attempt, from the very first beginning, agriculture and PV plants had difficulty in combination. This is because, for PV energy, there is a major need for commitment of significant areas of land to produce sufficient electricity, which, on the other hand, reserves a large area of land that could be used for cultivation. Facing this conflict, it is found urgent to propose effective solutions that would allow these two independent sectors to coexist with minimum disaffects and also find synergies and common points for their convergence. This has been considered a major topic of consideration since the last decades when the energy crisis, global warming, increasing hunger and clean water shortage were determined as important challenges for the sustainable development of humankind. Under these challenges, the agrivoltaics concept emerged.

Agrivoltaics technology has emerged as one of the most promising solutions to alleviate the land-use competition between energy and food production, as well as to increase land efficiency (see Fig. 3.1). Agrivoltaics or agro-photovoltaics is a concept that was originally conceived by Adolf Goetzberger and Armin Zastrow in 1981, with Germany being one of the countries that first began to realize the problem of land shortage for energy and food production [2]. These two men first theorized the concept of dual usage of arable land for solar energy production and plant cultivation, which would address the problem of competition for the use of arable land between solar energy and crops production. Recognizing this concept, Akira Nagashima developed the first prototypes of PVs for agriculture in Japan in 2004 (referred to as "solar sharing"), studying different shadowing rates in cultivation based on the concept of light saturation point of each crop (plants exploit only a small percentage of incident sunlight to photosynthesize), with the excess of solar light to be used by PVs to produce electricity [3]. However, it was in 2010 when the first experimental pilot project took place, with the first agrivoltaic systems being installed in France, and the term "agrivoltaic" being coined in 2011 [4]. Today, "agrivoltaics" term is widely used to refer to the approach of simultaneous use of areas of land for both PV power production and cultivation. Agrivoltaic practices are increasingly applied nowadays as emerging "single-step" solutions of the Water-Energy-Food-Ecosystems (WEFE) nexus challenges due to the multiple interactions with the WEFE resources. The emergence of this concept is assisted by the great decrease in the PV energy cost under rapid technological improvements.

For the evaluation of an agrivoltaic system, today several types of metrics are used. Initially, since an agrivoltaic system is composed of PV modules and farmland, the overall system viability is usually defined by the metric land equivalent ratio (LER) that allows comparing the conventional approach (farm and PVs set up separately) with the integrated agrivoltaic solution that essentially exploits the same land area [4, 6, 7]. Namely, LER is a measure of whether the combined value of agricultural and electricity production is equivalent to or greater than the value of singular use of land. LER is defined using Eq. 3.1, where $Y_{APVcrop}$ refers to the agricultural yield (e.g., kg/ha) under the agrivoltaic approach, Y_{crop} stands for agricultural yield in conventional farming, Y_{APVe} refers to the electricity production using the agrivoltaic

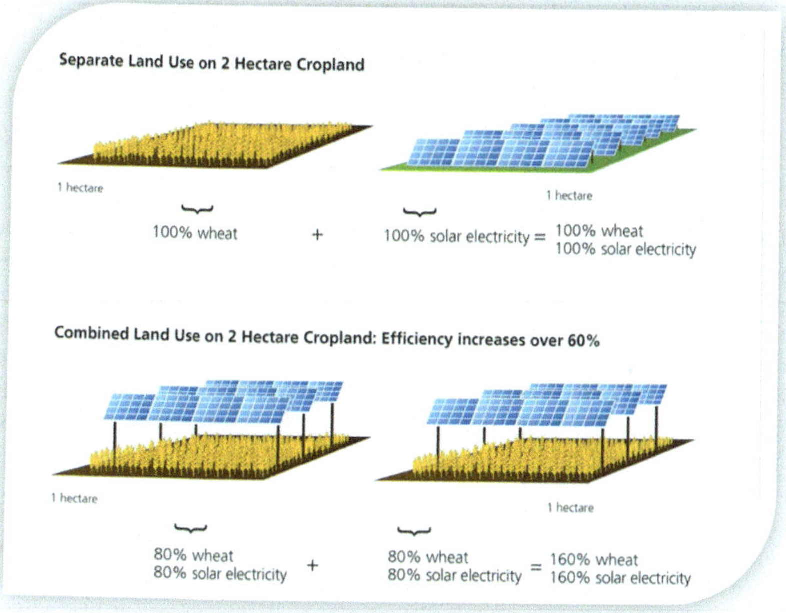

Fig. 3.1 Land efficiency comparison between traditional farming and agrivoltaics approach [5]

system and Y_{PVe} refers to the electricity production using a standard PV plant. LER values above 1 indicate that the agrivoltaic system is more effective than separate conventional PVs installation and farming for the same land area. At this point, it is noteworthy to mention that LER should only be used as a performance indicator comparing agrivoltaic systems for similar climatic conditions, PV system technology and crop type.

$$\mathrm{LER} = \frac{Y_{\mathrm{APVcrop}}}{Y_{\mathrm{crop}}} + \frac{Y_{\mathrm{APVe}}}{Y_{\mathrm{PVe}}} \tag{3.1}$$

Besides, the ground coverage ratio (GCE) is an additional performance indicator that can be used to specify the performance of an agrivoltaic system. GCE is defined as the ratio between the area of the PV modules and the area of land that they occupy (expressed in percentage). High GCE values usually mean a high energy yield due to the amount of solar radiation that reaches the PVs, whereas the crop yield will be low.

On the other hand, the impact of PVs on agricultural production can also be quantified using the relative water usage efficiency (WUE_{rel}) factor, as described in Eq. 3.2, where WUE_{APV} refers to the water efficiency under agrivoltaics approach and WUE_{crop} refers to the corresponding term under conventional farming approach

(WUE is commonly expressed as a unit of biomass per unit of water used in kg/m^3) [7]. WUE can be used as a measure of the amount of biomass produced per unit of water used by the plant, and it is a useful parameter to assess the benefits of the WEFE nexus, especially in drylands.

$$\mathrm{WUE_{rel}} = \frac{\mathrm{WUE_{APV}} - \mathrm{WUE_{crop}}}{\mathrm{WUE_{crop}}} \tag{3.2}$$

Under the agrivoltaics approach, there are a variety of fascinating benefits that are now explored. Some important ones are: (1) more land for renewable energy sources; (2) increment in total revenue of the land-owners; (3) beneficial cultivation under PVs shading (e.g., plants protection against high solar radiation and other extreme weather conditions, reduction of wind erosion); (4) increased water efficiency by exploitation of the water used for PVs maintenance (e.g., cleaning) for subsequent irrigation proposes; (5) better microclimate for PVs operation by the decrease of the ambient air temperature when crops are cultivated beneath PV systems; (6) new job opportunities; etc. Namely, the agrivoltaics approach can bring unique synergetic effects in favor of the WEFE nexus. Through a number of qualitative analyses carried out up until today, it is found that agrivoltaic systems can achieve a positive impact on more than 14 out of 17 SDGs set by the United Nations [8, 9].

Due to the aforementioned benefits, the agrivoltaics approach has nowadays attracted increasing attention from both farmers and energy suppliers, with their global installed capacity presenting exponential growth in the last decade, from around 5 MW_p in 2012 to higher than 14 GW_p in 2021 (see Fig. 3.2). The potential of agrivoltaics to boost a sustainable WEFE transition is significant by considering that the global energy demand could be offset even if 1% of croplands are converted into an agrivoltaic approach [11]. This potential is considered even more ideal for application in the Mediterranean area, where cultivation area and annual solar resource are quite high ($\approx$200 million ha and >1800 kWh/m^2, respectively) [12, 13].

Nevertheless, the development of the most convenient strategies to implement this technological solution to improve the socio-economic adaptation according to the WEFE challenges is still pending. Although many countries have demonstrated the technical and economic feasibility of agrivoltaics, there are still obstacles preventing their widespread exploitation. These include legislation issues for their approval for their construction and lack of adequate funding. Public backing for agrivoltaics is another hurdle for this concept in some regions. On the other hand, from the technology point of view, agrivoltaic strategies require a higher level of maturity, mainly in the part of energy production by the PVs and its subsequent usage, achieving simultaneously minimum disaffects for the near cultivations [10].

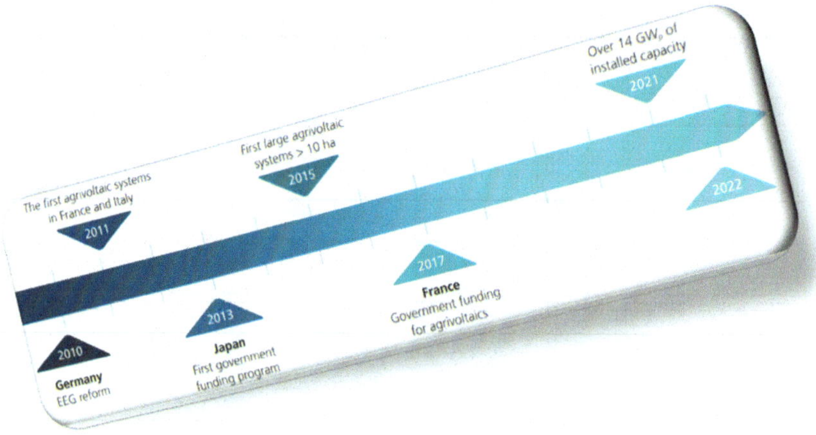

Fig. 3.2 A brief history of agrivoltaics since 2010 [10]

3.2 Application Modes of Photovoltaics in Agriculture

The goal of increasing renewable energy production and the associated incentives has led to a massive development of the PV industry. Today, light-to-electricity conversion using PVs is one of the most mature renewable energy production approaches. Regarding the agricultural sector, the technical solutions for integrating PV systems into farms are as diverse as farming itself. A kind of classification of agrivoltaic systems and some corresponding examples are presented in Fig. 3.3.

Agrivoltaics can be broadly categorized into open and closed systems. Open systems can be divided into ground-level overhead and interspace PVs, while closed systems are mainly referred to PV greenhouses. Regarding the first category, overhead systems are usually mounted at least 2 m above the ground in order for the land beneath the PV modules to be used for farming. Whereas, in the case of interspace systems, the land between the PV modules is the place that can be used for cultivation. On the other hand, PV greenhouses is an approach that is used to marry electricity production with increased agricultural activity, where the greenhouse improves the yield and quality of the crops produced inside in terms of micro-climate optimization.

Focusing initially on the open agrivoltaic systems, some benefits arising from the overhead installations are mainly related to the use of the land more effectively, as well as that the crops can be protected by the PV modules against adverse environmental effects. On the other side, interspace systems are that they have lower costs and tend to impact the landscape less. Both types of systems can be developed either with a fixed substructure or with a solar tracking system that can allow for better sunlight exploitation as the PV modules can be tilted (single or dual axis tilting) toward the direction of the sun. Some common applications of open-field agrivoltaics are related to permanent grassland, arable farming, horticulture and aquaculture. In these cases, the farmer, except for electricity production, can achieve animal feeding, vegetables,

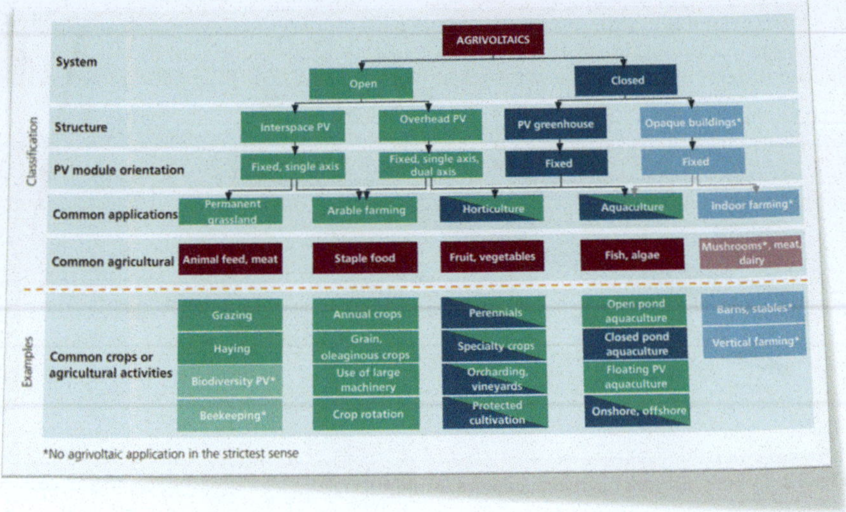

Fig. 3.3 Classification of agrivoltaic systems and some common examples [10]

fruits or other plants cultivation, as well as fish farming [10]. Of course, depending on the type of farming, the characteristics of the agrivoltaic systems have to be modified to achieve the minimum disaffects to the near cultivation. In this direction, institutes and universities are nowadays working to develop standards for the agrivoltaics field, where the requirements for agricultural use of PV modules are set. The standards aim to clearly differentiate agrivoltaic systems from traditional ground-mounted systems by outlining the criteria for primary agricultural use of agrivoltaics, which is expected to be a major prerequisite for ensuring agrivoltaic systems are successfully brought to market. A key premise that applies to all agrivoltaic approaches is that the land utilized for agrivoltaics must continue to be used for agricultural purposes [10]. As an example, Table 3.1 tabulates an overview of categories and forms of land use (open agrivoltaic systems) as set out in DIN SPEC 91,434 developed by Fraunhofer ISE, University of Hohenheim, German Institute for Standardization (DIN) and other partners in research and industry. An illustration of the categories and forms of land use as set out in DIN SPEC 91,434 is presented in Fig. 3.4.

According to this standard, the key requirements for the development of agrivoltaic systems are: (1) the agricultural usability of the area shall be preserved and the anticipated form of land use shall be specified in an agricultural usage proposal; (2) land loss after installing the PV system should not be higher than 10% of the total project area for category I and 15% for category II; (3) light availability and homogeneity, as well as water availability, should be evaluated and adjusted to meet the demands of agricultural products; (4) steps should be taken to prevent soil damage and erosion caused by PV system installation, soil anchoring or water runoff from the PV modules; (5) it should be ensured that the agricultural production following the installation of the agrivoltaic systems to be at least 66% of the reference yield.

Table 3.1 Overview of categories and forms of land use as set out in DIN SPEC 91,434 [10]

Agrivoltaic system	Use	Examples
Overhead agrivoltaics approach (farming under the agrivoltaic system)	1A: Permanent and perennial crops	Fruits, viticulture, berries, hops
	1B: Single-year and long-term crops	Arable crops, alternating grassland, vegetables, fodder
	1C: Permanent grassland with mowing	Extensive and intensive commercial grassland
	1D: Permanent grassland with pasture	Permanent pasture, pasture rotation (e.g., cattle, sheep, goats, pigs and poultry)
Interspace agrivoltaics approach (farming between the agrivoltaic system)	2A: Permanent and perennial crops	Fruits, viticulture, berries, hops
	2B: Single-year and long-term crops	Arable crops, alternating grassland, vegetables, fodder
	2C: Permanent grassland with mowing	Extensive and intensive commercial grassland
	2D: Permanent grassland with pasture	Permanent pasture, pasture rotation (e.g., cattle, sheep, goats, pigs and poultry)

The reference yield is computed using either the three-year average of yields from the same agricultural field or equivalent data from relevant publications [10].

On the other side, there is an increasing interest in the approach of PV greenhouses nowadays. The implementation of PVs into the shell of a greenhouse has shown numerous benefits for a farmer. Of course, the design of the infrastructure should be regulated according to the cultivation being done. The application of PVs on the roof area, even on the sidewalls of the greenhouse, decreases the portion of solar light that is available inside the infrastructure and subsequently affects the development of the crop. Today, various approaches have been proposed for the application of PV devices in greenhouses, such as (a) partial shading with the usage of opaque solar cells or modules; (b) focusing the direct sunlight by Fresnel lenses; (c) semi-transparent modules; with the first approach being the most widely used up until now (see Fig. 3.5). For this case, the most common PV technology that is applied is the crystalline silicon (c-Si) solar cells, which can provide high performance and stability in light-to-electricity conversion with a low electricity cost. For a low impact on crop growth, the coverage of the greenhouse roof by the opaque PVs is usually determined at about 20%. On the other hand, alternative technologies are also tested, such as commercial thin-film solar cells or emerging technologies that could provide multiple additional benefits by exploiting their unique characteristics [14].

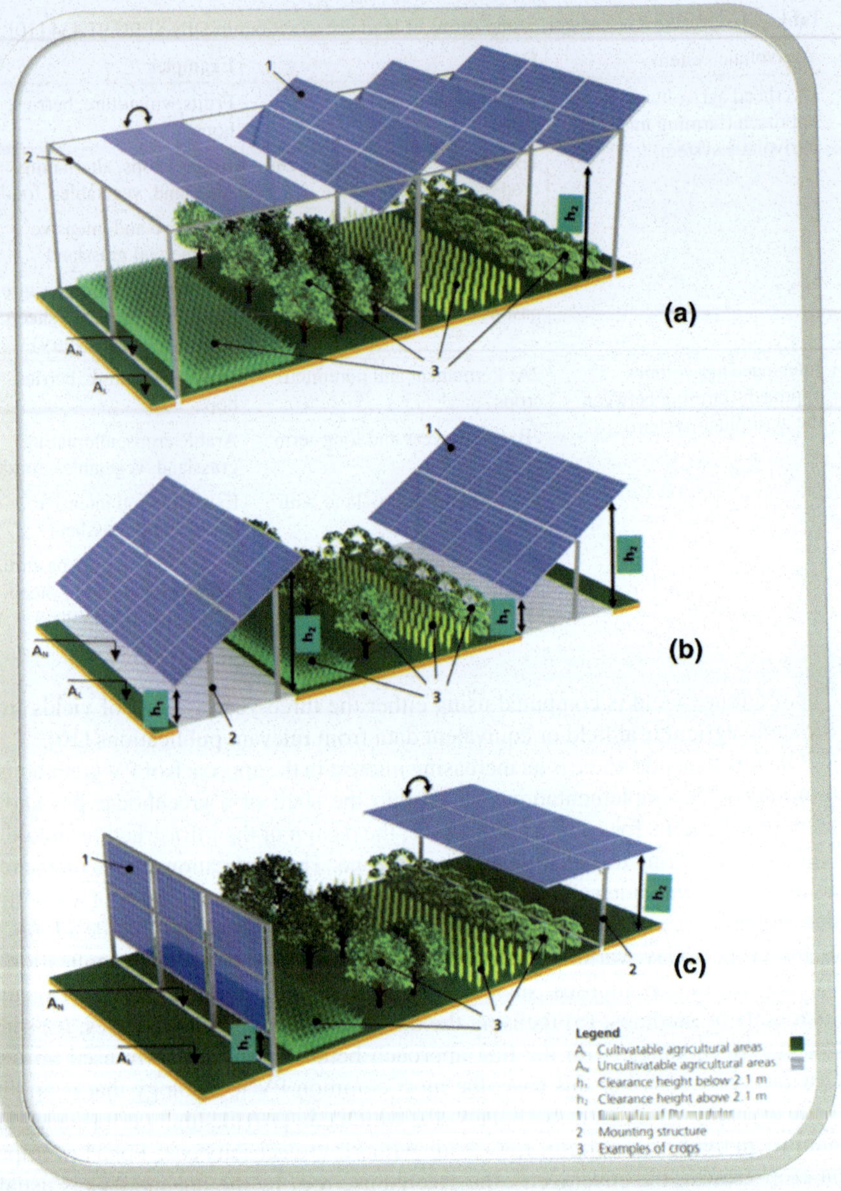

Fig. 3.4 Illustration of the categories and forms of land use as set out in DIN SPEC 91,434 [10]

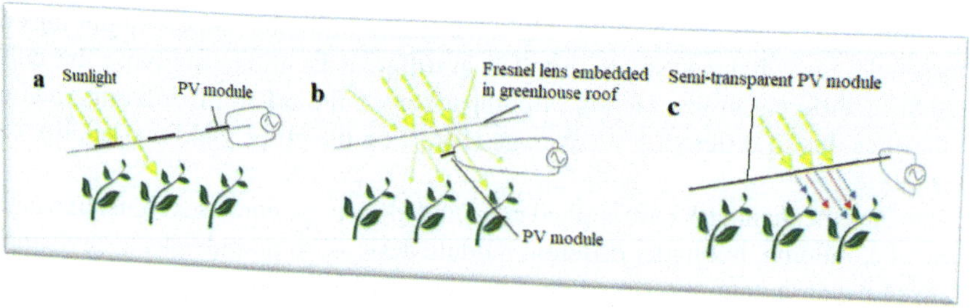

Fig. 3.5 Approaches that are explored today for photovoltaics application to greenhouses: **a** partial shading with opaque solar cells and modules, **b** focusing the direct sunlight by Fresnel lenses, **c** semi-transparent photovoltaics [15]

3.3 Case Studies on the Impact of Agrivoltaics on Crops Yield and Quality

It is a fact that today, there is a great number of demonstrative projects worldwide focused on agrivoltaic approaches and the impact of PVs installation in agriculture. The experience with varied design solutions suitable for upscaling to commercial scale is being gathered today at a rapid pace, with attention to be paid to both performance parameters for energy and food production, as well as to ecological impacts associated with specific design choices, particularly those related to landscape transformation issues. Most of them regard agrivoltaic systems based on the mature c-Si solar cell technology, with solar modules being developed under different designs, with the main focus being set on the shading effect and the subsequent adverse impact on the plants growth.

The extent of reduction in solar radiation under an agrivoltaic system and subsequently on cultivation growth depends on many factors, including seasonal solar altitude, the position of the plants underneath the PV arrays and the technical implementation of the facility. The latter includes orientation, tilt angle and size of the panels, as well as the distance between them. At the same time, the extent of the reduction of crops growth depends on the crop type and the stage of crop development when the shading is applied. The latter is demonstrated by various studies carried out in the past years, including cases related to wheat, rice and maize crops. However, there are not all cases that the shading effect has an adverse impact on cultivation. Some crops show increased yields under shading, especially in regions with high solar irradiation. In this case, shading can enhance the plant survival rate. This can also take place due to the potential effects of the agrivoltaic systems on farm microclimate (e.g., changes in evapotranspiration). In addition to yield factors, shading can influence the quality of the harvestable product, such as the protein, gluten and fat content. Besides, beneficial shading effects on cultivation growth can be associated with enhanced protection of crops against adverse climatic conditions, such as high temperatures and excessive rainfall.

Table 3.2 tabulates representatively some case studies and their main findings on open-field agrivoltaics systems installed in different locations, studying the cultivation of different crops. Herein, a comparison of the effect of different shading intensities due to different PV coverage (PV_R), on the crops yield and quality can take place.

On the other hand, PVs are applied more and more in greenhouses roofs, like in the case of a building, but under different module designs, to provide electricity supply or even power autonomy. Based on a number of studies on PV greenhouses, PV_R of the greenhouse roof lower than 20% has almost zero/negligible adverse effect on crops growth (of course, this number is indicative and depends on the climate where the greenhouse is located, type of crops and orientation of the greenhouse). On the other hand, several reports on PV greenhouses were found using more than 20% PV_R. Table 3.3 tabulates representatively some case studies on PV greenhouses employing c-Si solar cells, developed at different locations around the world, considering different greenhouse types and sizes, crop types and PV_R, demonstrating for each case the main findings.

Table 3.2 Case studies focused on open-field agrivoltaic systems and their corresponding main findings on the effect of photovoltaics on crops growth

Location	Agrivoltaics type/size	Crop type	PV_R (%)	Main findings	References
Montpellier, France (Latitude: 43° 15′ N, Longitude: 3° 87′ E)	Overhead, 4 m above-ground, 6.4 m × 6.4 m grid/860 m^2	Lettuce, durum wheat, cucumber	0–50	Reduced growth rate during the juvenile phase of the crops. No significant effect during the period of maximal vegetative growth	Marrou et al. [16]
Herdwangen-Schönach, Germany (Latitude: 47° 85′ N, Longitude: 9° 14′ E)	Overhead, 5 m above-ground, row distance of 3.4 m/3000 m^2	Winter wheat, potato, grass-clover, celeriac	30	In the first year, the crop yield losses were determined between 5 and 20%. In the hot and dry summer of the next year, the growth of winter wheat and potatoes benefited	Weselek et al. [17]
Southern Germany	Overhead, 5.5 m above-ground, row distance of 9.5 m/1206 m^2	Potato, winter wheat, celeriac, clover grass	<40	Decreased yield of the crops by up to 26%, mainly for potato and winter wheat. The agrivoltaics approach is advantageous under dry and warm climatic conditions	Trommsdorff et al. [18]

(continued)

Table 3.2 (continued)

Location	Agrivoltaics type/size	Crop type	PV_R (%)	Main findings	References
Illasi (Verona province), Italy (Latitude: 45° 28′ N, Longitude: 11° 11′ E)	Overhead, 1.9–2.6 m above the soil	Corvina grapevines	75	The agrivoltaics approach benefited the metabolic activity of vines in the midday. Vine yield was reduced more as cluster weight than berry weight, whereas pruning weight was almost the same	Ferrara et al. [19]
Mace Blvd, Davis, USA (Latitude: 38° 53′ N, Longitude: −121° 69′ E)	Overhead, 2.1 m above ground	Kale, chard, broccoli, peppers, tomatoes and spinach	25–93	Crops like kale, chard and tomatoes can be planted under the solar array with only a limited yield penalty as long as light levels are at least 55% of full-sun irradiance. Other crops like spinach are not advisable for farming under solar panel arrays due to their strong dependence on solar irradiance for the best crop yields	Hudelson and Lieth [20]
Daegu, South Korea (Latitude: 36° 14′ N, Longitude: 128° 78′ E)	Overhead, 3.3 m above the soil, panels distance of 1.5 m/1000 m^2	Rice, soybean and adzuki bean, onion, garlic, rye, monocropping forage corn, mixed planting corn with soybean	30	No negative impact of the agrivoltaic system on the forage yield of rye and corn, while rice yield was significantly reduced	Jo et al. [21]
Pujiang, China (Latitude: 30° 19′ N, Longitude: 103° 25′ E)	Overhead, 3 m above-ground, 4 m × 5 m grid/ 360 m^2	Kiwifruit	19–38	The kiwifruit volume and yield were slightly reduced in the case of 19% shading. The negative effects of shading on crops growth were significant for shading >30%	Jiang et al. [22]

Table 3.3 Case studies focused on photovoltaic greenhouses employing c-Si solar cells and their corresponding main findings on the effect of photovoltaics on crops growth

Location	Greenhouse type/size (m^2)	Crop type	PV_R (%)	Main findings	References
Pyrgos, Greece (Latitude: 37° 79′ N, Longitude: 21° 35′ E)	Small-scale experimental/ 4.26	Lettuce	20	No significant effect on plants growth	Trypanagnostopoulos et al. [23]
Sardinia, Italy (Latitude: 39° 19′ N, Longitude: 8° 59′ E) (Latitude: 39° 22′ N, Longitude: 8° 55′ E) (Latitude: 40° 38′ N, Longitude: 8° 39′ E)	Gable roof/ 480 Venlo/840 Mono-pitch/ 837	14 horticultural and floricultural crops low light: dracaena, lanchoe, poinsettia medium light: ficus, chrysantemum, rose, asparangus, spinach, basil, strawberry, lettuce high light: tomato, cucumber, sweet pepper	25–100	Limited yield reductions for all considered crops for PV_R of 25%. Medium-light demanding horticultural species can be grown even with PV_R up to 60%. Low-light demanding crops can be cultivated with negligible or limited yield reduction, even by a PV_R of 100%	Cossu et al. [24]
Almeria, Spain (Latitude: 36° 53′ N, Longitude: 2° 22′ W)	Venlo/2415	Tomato	15–50	Moderate shading did not affect crop growth. PV_R over 30% decreased significantly crops yield and quality	López-Díaz et al. [25]
Agadir, Morocco (Latitude: 30° 13′ N, Longitude: 9° 23′ W)	Canarian/172	Tomato	40	The PV panels protect the crop from intense solar radiation in summer as they play the role of shading screens. No significant effect on the microclimate and tomato yield in winter	Ezzaeri et al. [26]
Kunming, China (Latitude: 25° 07′ N, Longitude: 102° 68′ E)	Small-scale experimental/ 26.25	Strawberry	25.9	Improvement in the yield and quality of fruits	Tang et al. [27]
Shiraz, Iran (Latitude: 29° 21′ N, Longitude: 52° 31′ E)	Flat arch/ 4081	Rose	Dynamic, up to 22.4	The optimum PV_R was found at 19.2%. Illumination required by the plants was perfectly provided, while the energy demands and CO_2 emissions were greatly reduced	Alinejad et al. [28]

3.4 Costs for Photovoltaics-Farming Dual Land-Use

Adopting PVs in agricultural land provides mutual benefits toward sustainable WEFE management, including an increase in land-use efficiency. However, there is presently little information available on the economic efficiency of this strategy, with the statistics showing increased costs arising from the planning, design and development of agrivoltaics, as well as the coordination across additional stakeholders. Understanding the capital expenses of agrivoltaics is merely the first step in determining the economic viability of this strategy. PVs efficiency and ground coverage ratio are important cost drivers for these systems, particularly for cases of simultaneous PVs and crops applications, which have higher balance-of-systems costs. Additional data on crop yield changes, water use efficiency impacts, ground cover effects on PVs system temperature, as well as operational and maintenance costs, are required to assess the lifetime cost and revenue effects of PVs-farming dual land-use to truly understand the value proposition of this strategy under various scenarios [29].

According to previous studies, agrivoltaic project developers may incur higher costs from the approval stage process. These expenses include the identification of the location, soil and environmental assessments, development plans, as well as legal fees. Ground preparation, on the other hand, can be an additional cost concern for guaranteeing trouble-free agricultural activities. Agrivoltaic plants may also have to be built with more specialized technology, such as using trackers, bifacial panels and semi-transparent PVs, which result, along with the more complex mounting systems, in higher costs [30].

Indicatively, the investment costs for an overhead and interspace agrivoltaic system, compared to a conventional PV ground-mounted system, are presented in Table 3.4. The given values are only intended to give a rough overview of the proportions of the individual cost items and enable an approximate comparison of the economic efficiency of different system designs. The exact costs depend, among others, on the size of the system, the location, the distance to the grid connection point and the current developments on the PVs market.

Site preparation costs also contribute to differences in the total costs across scenarios. Depending on the type of dual land-use, such as PVs + cultivation or

Table 3.4 Investment costs of different photovoltaic system types (in euros per kWp) [31]

Cost point	Conventional PVs	Overhead agrivoltaic	Interspace agrivoltaic
Planning and development	85	300	53
Module	220	360	252
Mounting system	75	400	191
Power connection	174	174	174
Fencing	18	0	18
Total	572	1234	688

PV + grazing versus baseline PV structures, the capital, operational and maintenance costs are different. Cultivation and grazing incur some additional costs, such as for fencing or a water well for grazing animals, while between the aforementioned cases, fencing costs are higher for the PVs + cultivation scenario because farmers should achieve extra fencing to protect the crops from wildlife. Usually, all dual-use PV approaches have also increased site investigation costs compared to typical bare ground-mounted PVs, owing to additional planning, analysis and coordination needed to ensure the success of the vegetation under the panels. For example, this includes working with a manager who coordinates between developers, regulators and farmers, investigating the implications of panel shading on crops or grasses and verifying compliance with dual-use incentive program requirements. On the other hand, compared with conventional PV mounting, the PV + cultivation and PV + grazing scenarios reduce some site preparation costs, including costs related to grubbing and clearing. Taking into account that for both aforementioned scenarios the land was previously farmland, grading costs are also lower [29].

On the other hand, the economics of PV greenhouses are gaining increasing attention in the last few years. According to previous reports, PV greenhouses can achieve good economic performance, providing, in many cases, stable PV power sale revenues and high-quality crop sale revenues. The annual return on investment (AROI) and internal rate of return (IRR) vary depending mainly on the crops produced in the PV greenhouses, with their values reaching even 20%. The discounted payment period (DPP) is usually determined from 4 to 8 years, which is shorter than conventional greenhouses. This makes it easier for PV greenhouse owners to find finance [32]. Figure 3.6 presents an illustration of five different PV greenhouse systems installed in China, as provided in a previous investigation. The corresponding information and the economic performance of these systems are given as an example in Table 3.5.

3.5 Current Status in the Agrivoltaics Market

In 2021, again, solar power was ranked the top power source installed across the globe, with PV installations presenting the fastest growth amongst the renewables, with a 21-fold increase in the 2010–2021 period. By the end of 2021, the cumulative installed capacity of PVs reached 843 GW globally, approaching the largest renewable power source in terms of installed capacity, which is hydropower. Only the last year, 133 GW of PVs capacity was commissioned, with 57% of this number to regard installations made in Asia [33]. From the financial point of view, the global PV market was estimated at USD 159.84 billion in 2021, and it is predicted to hit over USD 250.63 billion by 2030, poised to grow at a compound annual growth rate (CAGR) of 5.1% during the forecast period 2022 to 2030. This is a result mainly arising from the major cost reductions backed by technological advancements, high learning rates, policy schemes and incentives, and innovative financing models [34].

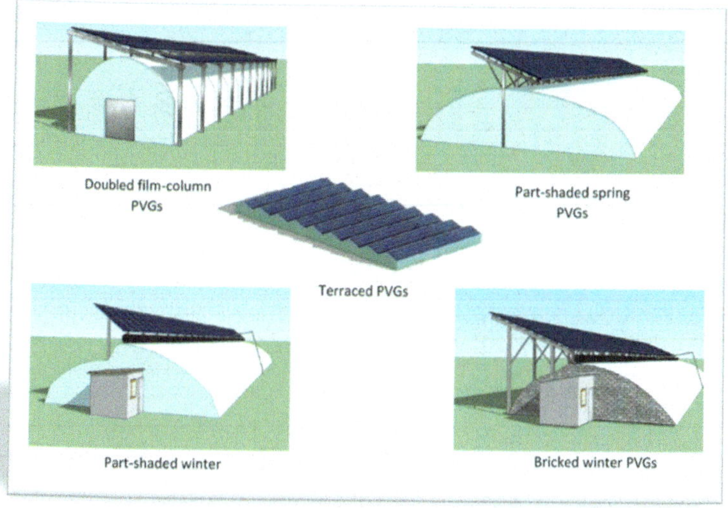

Fig. 3.6 Illustration of five different photovoltaic greenhouse systems [32]

As a part of the rising PVs market and under the pressure of WEFE challenges, the agrivoltaics market is gaining immense interest in our days (see Fig. 3.7), becoming a multi-billion dollar industry in the last few years. More specifically, in 2021, the global agrivoltaics market was valued at USD 3.17 billion. To the growth of the agrivoltaics market, determinant factors were the advantages of agrivoltaics integration in modern farmlands and the rising awareness of this strategy, the reduced prices of light-to-electricity conversion by the PV systems and the subsequent high-level satisfaction of the customers as a result of their increased profits, the improvements in government strategic planning and initiatives for profits enhancement, as well as the rising investments in research and development of novel agrivoltaic systems that can bring new horizons in solar farming in terms of efficiency [37].

Based on the system design, the agrivoltaics market is now divided into fixed and dynamic agrivoltaics. From them, the fixed systems cate to significant demand in this market. They are the most applied systems above open-field crops or between crops, or on greenhouses, while they are optimized by the regulation of the density and inclination of the solar panels depending on the crop type. On the other hand, another classification of the agrivoltaic market is according to the crop type, namely root crops, leafy vegetables and fruits. From those, a substantial growth rate over the next period is expected to show the green leaf segment since these crops require limited sunlight to grow, as is usually the case under an agrivoltaic system. As for the major players operating in the agrivoltaics market today, some of them are Agrivoltaic Solution LLC, Sun Agri, REM TEC, Enel Green Power, Boralex, BayWa AG, TotalEnergies SE, Mackin Energy, Sunrise Power Solutions and Suntech Power Holdings [38].

Table 3.5 Information on the five different photovoltaic greenhouses and their corresponding economic performance [32]

PV greenhouse		Terraced	Doubled film-column	Part-shaded spring		Part-shaded winter		Bricked winter
Size (m^2)		3588	360	540 unshaded	360 shaded	594 unshaded	360 shaded	994
Suitable crops		Tea tree	Edible mushrooms	Vegetables	Edible mushrooms	Off-season vegetables	Edible mushrooms	Off-season vegetables
Installed capacity (KW)		156	56	51		51		84
Investment	Greenhouse	550	125	110		260		220
	PV	[illegible]	420	375		375		630
	Equipment	15	30	5		10		12
AROI (%)		9.76	13.92	15.16		20.71		13.03
IRR (%)		12.05	16.28	17.48		21.89		15.27
DPP (%)		8.00	5.89	5.53		4.63		7.26

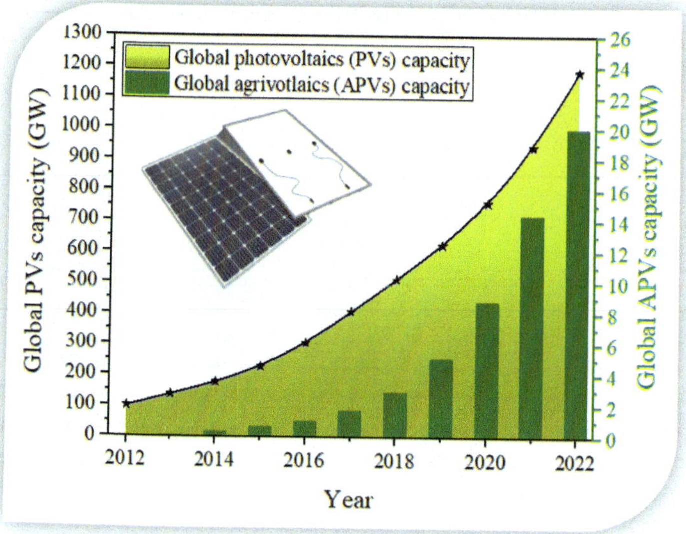

Fig. 3.7 Solar photovoltaics and agrivoltaics capacity by year. (Reproduced using data taken from references [35, 36]. The photovoltaic panel 3D model was taken from sketchfab, "Photovoltaic Panels" (https://skfb.ly/oyt86) by mfb64, licensed under Creative Commons Attribution (http://creativecommons.org/licenses/by/4.0/)

The agrivoltaics market share by region in 2021 is presented in Fig. 3.8. According to market reports, North America gained the highest market share as a result of a number of technological interventions in agricultural processing. Besides, this was because of the development of advanced agricultural machinery and equipment in this region. On the other hand, Europe has accounted for the highest share of the agrivoltaics market due to the major scarcity of farmlands, with the rise of the agrivoltaics approach aiming to increase land-usage efficiency for both energy and food production. After Europe, Asia was one of the largest market holders for agrivoltaics and is expected to be one of the fastest developing sectors in terms of CAGR growth throughout the next years [37]. Considering the global PVs market, in the next years, China will be the world's top solar agrivoltaic system producer and home of the biggest installation projects. In 2021, solar greenhouses in China covered up to 1.96 million hectares or 30% of the total horticultural areas in China. The most popular vegetables grown in the solar greenhouses include tomatoes, cucumbers, peppers, strawberries, melons, eggplants and leafy vegetables [36].

Regarding the future of the agrivoltaics market, by the next decade, it is foreseen to reach around USD 8.9 billion, with the market analyzers estimating a CAGR of 12.15% from 2022 to 2030 (see Fig. 3.9).

The future potential of the agrivoltaics market will greatly depend on greenhouse installers, which are the main clients. The global greenhouse market size accounted

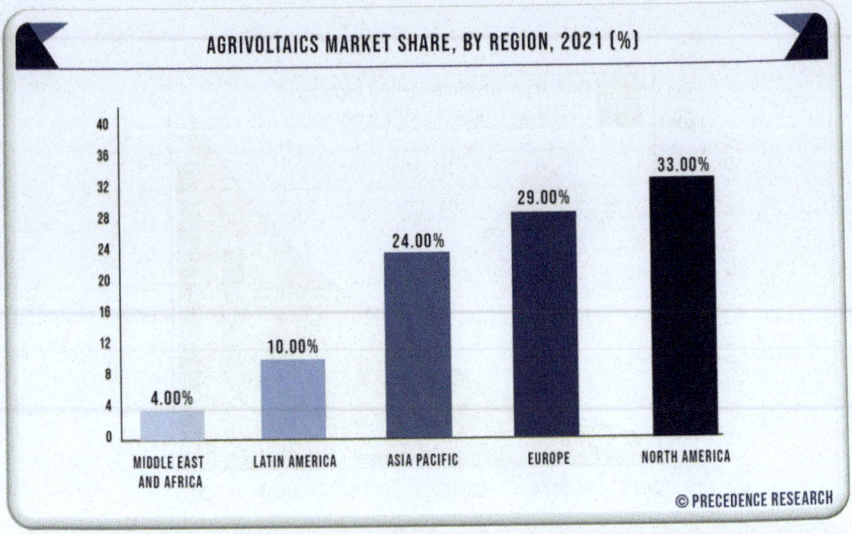

Fig. 3.8 Agrivoltaics market share by region for the year 2021 [37]

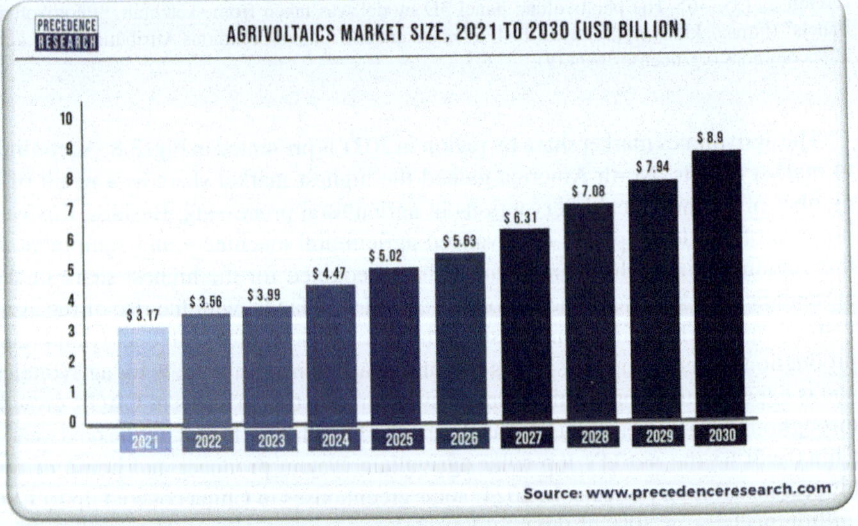

Fig. 3.9 A forecast on agrivoltaics market size until 2030 [37]

for USD 34.8 billion in 2021 and is expected to hit around USD 78.9 billion by 2030, poised to grow at a compound annual growth rate CAGR of 9.52% from 2022 to 2030. The commercial greenhouse market share by Region, in 2021, was 32.8% for North America, 29% for Europe, 24.4% for Asia Pacific, 9% for Latin America and 4.8% for the Middle East and Africa [39]. From the aforementioned, Asia–Pacific is seeing the quickest growth. The dominant player by far nowadays in the greenhouses market in terms of land covered by greenhouses is China, with 2.76 million ha of greenhouses deployed [40]. Today, Chinese greenhouses are considered the lowest-cost, and by considering the ultralow-cost of PVs manufacturing in this Country, China will take a leading position in the agrivoltaics market in the next years. Of the 2.8 GW agrivoltaics systems installed globally, China had roughly 1.9 GW of capacity as of 2020 [36]. In Europe, Spain is leading in protected agriculture with 52,170 ha, mostly under low-cost polyhouses. A table showing the worldwide total area in major greenhouse production is shown in Fig. 3.10 [40]. The dominant type of greenhouses for 2021 was the plastic, which accounted for 59% of the commercial greenhouse market share, while plastic greenhouses is predicted to grow at a higher rate than that of glass greenhouses in the next years. Some of the key market players in the greenhouses market are Berry Global, LumiGrow and Agra-tech, Inc for the US, Signify Holding, Heliospectra AB and Plastika Kritis for Europe, and Everlight Electronics, Top Greenhouses and Saveer Biotech Ltd for Asia [39].

However, the agrivoltaics market still faces some key challenges that limit its growth rate. Amongst them, at first, it is the lack of agrivoltaics approach awareness. This is because the modern agricultural equipment introduced by key market players is more expensive than the standard types of machinery available in the market. Combined with the lower disposable income of farmers and the difficulty in obtaining funding for the installation of agrivoltaic systems, these make it difficult for farmers to adopt such modern technologies, especially in underdeveloped countries. Besides,

Country	*Greenhouse area (ha)*
China	2,760,000
Korea	57,444
Spain	52,170
Japan	49,049
Turkey	33,515
India	30,000
Italy	26,500
Mexico	11,759
Netherlands	10,370
France	9,620
USA	8,425

Fig. 3.10 Greenhouse manufacturers around the world [41] (Reproduced from Ref. [41], <a href = "https://www.freepik.com/free-photo/couple-workers-standing-garden-near-flowers-talking_7286290.htm#from_view=detail_alsolike" Iamge by drobotdean </a> on Freepik, free license after attribution)

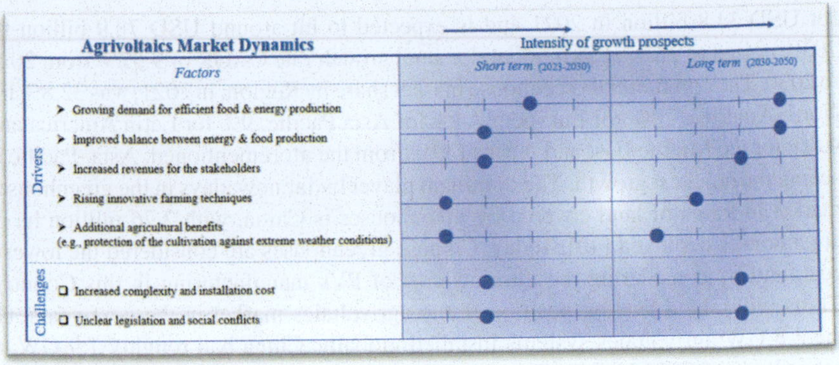

Fig. 3.11 Analysis of the agrivoltaics market dynamics

the growth of the agrivoltaics market is hampered to a great extent by the lack of legalities and skilled workforce, as well as due to social conflicts [42]. An analysis of the agrivoltaics market dynamics, presenting the key drivers and challenges in the field, is presented in Fig. 3.11.

3.6 Significance of Photovoltaic Agriculture in the Mediterranean Area

Around the Mediterranean Sea, more than twenty sovereign countries with similar climate features border the Mediterranean area. From them, the Northern Mediterranean region accounts for one of the world's fastest-growing areas in terms of economic, industrial and urbanization rates, consequently demonstrating highly increased energy and land demands. Parallel to the rapid growth, this area is also set at the center of climate discussions. According to recent research reports, the mean temperature in this region will increase by a significant rate in the next few years if the appropriate countermeasures to decrease greenhouse gas emissions are not taken. This would lead to changes in Mediterranean ecosystems that have been unmatched in the past 10,000 years [43].

Mediterranean energy demand surpassed 1022 megatonnes of oil equivalent (Mtoe) in 2018. In the next years, this value decreased since the region was hit hard by the COVID-19 pandemic, with the North Mediterranean countries being hit the hardest, mirroring the severity and length of lockdowns put in place. To this crisis, low-carbon fuels were proven the most resilient, with renewables increasing at an all-time high. However, the energy demand trend is expected to change by the time we pass, returning to pre-COVID-19 levels and showing a great increase in the next years. In the reference scenario (taking into account past trends, current policies and ongoing projects), the total energy demands are expected to reach 1404 Mtoe by

2050, poising a growth of 37%, setting except clean energy production, the energy efficiency and conservation substantial [44].

Amongst the power sources, renewables showed the highest growth in the Mediterranean region in the last decade. Power generation capacity from renewable energy technologies reached 686 GW in 2020. The leading region in terms of renewable energy systems installations is the North Mediterranean, with the South and East areas accounting for 20% of the total capacity. In pre-COVID-19 years, the investments in new renewable energy capacity were calculated at tens of billion USD per year, with the largest share to be concentrated in the North Mediterranean. Nonetheless, due to abundant resources, numerous South and East Mediterranean countries are now establishing themselves as particularly attractive markets for renewable energy technology. The outlook for 2050 indicates that renewables will supply 268 Mtoe in the reference scenario, almost doubling the output compared to 2020 levels, highly surpassing the power production from fossil fuels. Adopting energy transition strategies to achieve carbon neutrality by 2050 would result in a 2.6-fold increase in renewable energy consumption in North Mediterranean countries, i.e., from 91 Mtoe in 2018 to 243 Mtoe in 2050. On the other hand, the corresponding increase in the South Mediterranean countries is calculated at sixfold compared to 2018 levels, i.e., 167 Mtoe in 2050. As a larger deployment of renewables is taking place, a positive impact on socio-economic development will also be observed gradually [44].

From renewable power sources, solar PV energy is considered of the most attractive options, with the practical solar PV power potential in some areas of the Mediterranean to be determined at the highest levels across the globe (see Fig. 3.12). Today, PV energy is considered the third biggest source of renewable energy in the Mediterranean area, after hydropower and wind, with its further development to be limited by technical, financial and market barriers [45]. Table 3.6 tabulates the solar PV capacity by country in the Mediterranean region as presented in 2021, as well as the corresponding share of the total electricity consumption.

On the other hand, agriculture is the main land use, water consumer, and key activity for the Mediterranean's rural population. The ensemble of the Mediterranean countries has approximately 877 million ha of land, of which about 28% is devoted to agricultural activity, with the share of agricultural land demonstrating a pronounced spatial heterogeneity (from 4% of Egypt to almost 76% of Syria) (see Fig. 3.13a). Besides, significant are the differences in the farm structures and agro-management practices, both affected by the environmental conditions, rendering substantial variations in agricultural inputs (e.g., pesticides and water for irrigation) and outputs (e.g., crop types and yields). Annual crops in the Mediterranean area include mainly cereals (e.g., wheat, maize, barley and rice) and vegetables (e.g., potatoes and tomatoes), with their production rate to highly differentiate from region to region. Large spatial differences also characterize the livestock subsector, namely meat (beef and buffalo) and milk production [47]. A view of the production rate (2014–2018 average) of some of the main agricultural products in the Mediterranean area, as well as the percentage of agricultural land to the total land area by country, can be obtained from Table 3.7.

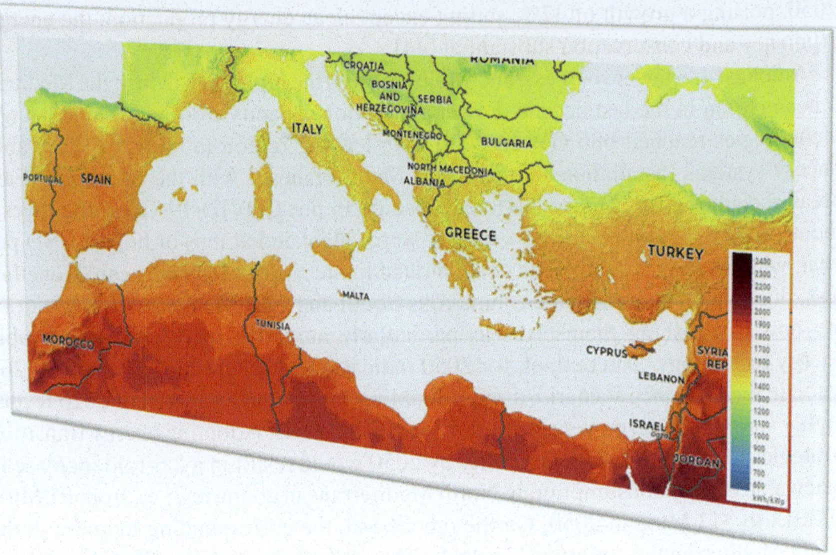

Fig. 3.12 The Mediterranean solar atlas as a part of the global solar atlas released in 2023 [13]. Data are obtained from the "Global Solar Atlas 2.0, a free, web-based application developed and operated by the company Solargis s.r.o. on behalf of the World Bank Group, utilizing Solargis data, with funding provided by the Energy Sector Management Assistance Program (ESMAP)

Table 3.6 Solar photovoltaic capacity by country and share of total electricity consumption in the Mediterranean area [46]

Country	Total solar PV capacity (MW)	Share of total consumption (%)
Algeria	448	
Bulgaria	1186	4.7 (2020)
Croatia	109	
Cyprus	316	3.3 (2016)
Egypt	1675	
Greece	3530	9.3 (2020)
France	14,718	2.8 (2020)
Israel	2555	4.7 (2019)
Italy	22,698	8.3 (2020)
Malta	196	6.5 (2017)
Portugal	1801	3.4 (2020)
Spain	15,952	9.0 (2020)
Turkey	7817	5.9 (2020)

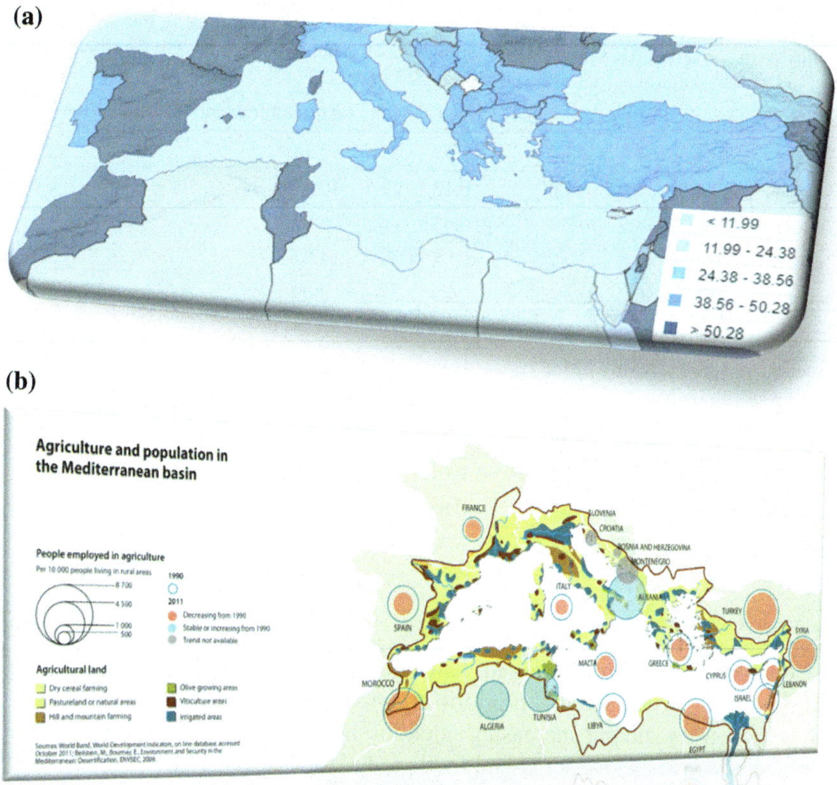

Fig. 3.13 **a** Percentage of agricultural land to the total land area by country and **b** a map showing people's employment in agricultural activities and the corresponding trend through time [48, 49]. (Figure 3.13a is reproduced from Ref. [48] (creator: THE WORLD BANK), under the terms of the CC-BY 4.0 license, Fig. 3.13b is reproduced from Ref. [49] (cartographer: GRID-Arendal), under the terms of the CC-BY-SA 4.0 license)

Despite the wide variety of agricultural products and their high production rates, many countries of the Mediterranean Basin still depend on food imports to meet the requirements of the increasing population. Besides, multi-faceted agri-food system challenges also appeared in the last years, with the Mediterranean region being widely acknowledged as one of the most exposed areas in the world to the effects of climate change, water scarcity, biodiversity loss and land degradation [50]. Simultaneously, as it is shown by many statistical data published in the last few years, there is a declining share of agricultural employment (see Fig. 3.13b). According to reports from the International Labor Organization, the percentage of people who work in agriculture has dropped from 44% in 1991 to 26% in 2020. This is partly down to the growing use of agricultural technology, but it also points to a bigger problem, which is that many people today do not want to work on farms anymore [51].

Table 3.7 Production average of cereals, fruit, vegetables, meat and milk in the Mediterranean countries for the 2014–2018 period in 10^5 tonnes and the percentage of land to the total land area reserved for agricultural activities by country (2020 statistics) [47]

Country	Cereal	Fruit	Vegetables	Meat	Milk	Agricultural land (% of total land area)
Algeria	40.4	67.6	63.4	1.57	35.8	17.36
Bulgaria	92.5	4.9	4.6	0.18	11.3	46.49
Croatia	30.6	3.2	12.4	0.43	6.8	26.89
Cyprus	0.4	1.8	0.7	0.05	2.5	14.52
Egypt	229.7	150.8	158.2	7.89	51.6	3.99
France	662.6	92.2	52	14.39	262.7	52.15
Greece	38.5	40.5	25.2	0.43	19.4	45.52
Israel	2.9	13.8	14.9	1.29	15.4	29.87
Italy	175.8	175.3	125.9	7.75	119.3	43.96
Malta	0.1	0.1	0.8	0.01	0.4	32.44
Portugal	11.9	19.6	24.4	0.89	20.8	42.28
Spain	211.6	192.7	128	6.32	80.2	52.33
Turkey	354.3	217.5	239.8	9.90	197.2	49.07

Considering the aforementioned challenges, and by taking into account that energy and food security are set as top priorities in several Mediterranean countries, agri-voltaics projects stirring increasing attention nowadays as a win–win strategy to combine effective clean-energy and food production, as well as to increase land-use efficiency.

3.7 Barriers and Opportunities for Further Development of Agrivoltaics

Although the technical and economic feasibility of agrivoltaics has already been demonstrated in a wide variety of cases and for different regions around the world, there are still multi-dimensional barriers preventing their widespread adoption. The acceptance of the agrivoltaics strategy is affected by a series of context factors. It is not only related to the applied technology and to the decisions of the interested parties but also to aspects that shape the context in which the process of creating acceptance takes place and that influence the perception of the issue from the outside by those who are being asked to accept it. These aspects include social, legal and environmental contexts and circumstances, as well as socio-political and normative conditions [10].

Current trends show that the growth of agrivoltaics is contingent on market adoption of the technology through community acceptance and supportive local regulatory environments. Themes related to their development, including complexity, risk of effectiveness and economic profitability, are frequently surfaced. According to the PV industry professionals, agrivoltaic projects are of increased complexity to be actualized, including added layers of intricacy in system design and increased coordination with stakeholders. Concerns range from technical details of accommodating a dual land-use, the impact, as well as other considerations, including the balancing of the interests of the stakeholders, all of which encumbering a project development. Herein, the prevailing emerging doctrine is: "You add something, it is more costly, more complex, needs more training, more people and stuff. At the end, it is harder to pull it off." Although the majority of people present the commendable benefits of agrivoltaics, at the time of the decision of project implementation, the extra effort needed for the development of a project dealing with a mixed-use power-agricultural site is effectively a deterrent. The economic profitability of an agrivoltaic project is also a common concern of the interested parties. Due to the risks and uncertainties of attaining an efficient dual land-use, investors express doubts about financing agrivoltaic projects. Similarly, participants also express worries about the expenses associated with the greater cooperation necessity to implement a dual-revenue stream, with money-saving to be considered from them, in many cases, better than profit [42].

If the agrivoltaics strategy is to succeed, it also has to be anchored in society through social acceptance. Even today, the development of ground-mounted PVs, including agrivoltaics, is resisted or rejected by some local societies, who, in many cases, criticize the location, form and size of these systems, expressing their fear that their recreational landscape will lose their value. Acceptance challenges arise at a local level when decisions are made under specific political or economic reasons to build new systems in a certain form and way, without adequately taking into account the concerns of the community, the interests of the population or giving local people the opportunity to voice their opinions. It is, therefore, no surprise when societal groups do not accept such decisions, even if they were made in accordance with legitimate planning and procedures. On the other hand, the unclear legislation on the steps that have to be taken to gain approval for agrivoltaics development in many cases is also a deterrent reason for adopting this strategy by the stakeholders [10]. Table 3.8 tabulates some of the commonly faced barriers and the corresponding recommendations to deal with them, as they are presented in previous discussions made in the field.

Although the aforementioned barriers are commonly faced before the implementation of an agrivoltaic project, the benefits arising from the strategic co-location of solar PVs and agricultural activity, and the opportunities that this approach offer, are many and cannot be omitted. Herein, a multi-dimensional assessment of the agrivoltaics projects should be considered in order to move forward to the widespread adoption of this strategy. Agrivoltaics projects do not only establish a dual-revenue stream for the involved parties. They are a substantial way to stimulate community acceptance toward a sustainable WEFE nexus management and establish better community partnerships for this aim. As shown by Wolsink's U-curve of local acceptance, the lowest levels of acceptability of renewable energy systems development are recorded during their siting phase. This indicates that the efforts to align projects with the community should be emphasized through project siting and planning stages. Herein, agrivoltaic

Table 3.8 Barriers and recommendations for further development of agrivoltaics [42]

Barrier	Explanation	Recommendation	Relevant actor
Complexity	Agrivoltaic projects are considered complex and require extra effort to actualize, including added layers of intricacy in system design and increased coordination with stakeholders	Provide financial incentives to photovoltaic enterprises pursuing agrivoltaics	– Photovoltaics developer – Local government – State government
Non-intimate local partnerships	Non-intimate relations between the potential stakeholders	Investment in stakeholder engagement and pursuit of meaningful partnerships to improve the development process	– Photovoltaic developer – Farmer – Local community
Safety and liability	Safety hazards to people and livestock and the potential for damage to electrical equipment are themes under consideration for developers and investors	Prior to development, design a contract between the involved stakeholders that protects against risks and establishes liability	– Photovoltaic developer – Farmer – Third-party insurer
Economic profitability	Unbalance in developing costs between photovoltaic developers and farmers	Development of mutually beneficial business models to support both parties financially	– Photovoltaic developer – Farmer
Local community acceptance	Resist or rejection of permitting projects of agrivoltaics in some local areas. Criticism for the location, form and size of agrivoltaics, expressing fear that the recreational landscapes will lose their value	Pursue development in a manner that purposefully upholds local values to enhance marketability and attitudes towards photovoltaic energy. Development of systems that are appropriate to be adopted in agriculture through incorporating existing agricultural practices	– Photovoltaic developer – Farmer – Local community – Local government
Policy	Unclear legislation on the steps that have to be taken to gain approval for agrivoltaics construction	Revise local bylaws to accommodate photovoltaics on farmlands. Develop state zoning enabling laws that explicitly preempt local solar restrictions in favor of agrivoltaic development	– Local government – State government – Policy makers

projects have the opportunity to move forward. Agrivoltaic projects that focus on preserving local identity and their design is sensitive to local nuances, interests and values can be effective in gaining community support. Community engagement and acceptance can be further achieved through designing agrivoltaic projects that view energy and food production as equally important, securing in this way sustainability. On the other hand, the development of agrivoltaic projects can be promoted in states that grant citizens involvement with defined rights throughout forums established by local governments, where they can express their support or opposition to a proposed system development and insist on specific outcomes. Moreover, financial incentives to explicitly encourage agrivoltaics strategy, as well as ease regulatory burdens for agrivoltaic deployments, can increase the interest in this strategy. Finally, agrivoltaics can represent a mean for communities to obtain energy sovereignty and can be coupled with initiatives for energy sovereignty, such as those policies that support community solar projects [42].

3.8 Outlook

To meet the global energy and food demands in this rapidly developing world, land-use efficiency has to be greatly increased, with the relationship between critical land uses aimed to be complementary rather than competitive. Herein, the agrivoltaics strategy is emerged as a sustainable approach to mitigate conflicts between solar and agricultural activities by providing mutual benefits and added value to each sector. Due to the great technological advancements in the PV industry, agrivoltaic systems can nowadays be developed in different application modes, providing high efficiency in double energy-food production. The number of case studies on the effect of PV energy exploitation combined with the different types of crop cultivation, for various latitudes and longitudes around the world, is rising at high rates. On the other hand, standards to build agrivoltaic systems are established, outlining the criteria and attributes that PV systems should meet for their integration into the agricultural sector. Under this trend, the global agrivoltaics market is now experiencing rapid growth, gaining billions of USD value, with the greatest installations to appear in regions with high practical solar PV power potential and agricultural activity. However, agrivoltaic projects still introduce unique challenges for both researchers and site operators, where novel technologies and further multi-dimensional assessment are needed to be highly considered, in order to establish agrivoltaics as a dominant strategy toward sustainability.

References

1. UN climate change conference UK 2021. https://ukcop26.org/. Accessed 5 Apr 2023
2. Goetzberger A, Zastrow A (1982) On the coexistence of solar-energy conversion and plant cultivation. Int J Sol Energy 1:55–69
3. Nagashima A (2005) Sunlight power generation system
4. Dupraz C, Marrou H, Talbot G et al (2011) Combining solar photovoltaic panels and food crops for optimising land use: towards new agrivoltaic schemes. Renew Energy 36:2725–2732
5. Fraunhofer ISE (2017) Harvesting the sun for power and produce—agrophotovoltaics increases the land use efficiency by over 60 percent. https://www.ise.fraunhofer.de/en/press-media/press-releases/2017/harvesting-the-sun-for-power-and-produce-agrophotovoltaics-increases-the-land-use-efficiency-by-over-60-percent.html. Accessed 5 Apr 2023
6. Mead R, Willey RW (1980) The concept of 'land equivalent ratio' and advantages in yields from intercropping. Exp Agric 16:217–228
7. Toledo C, Scognamiglio A (2021) Agrivoltaic systems design and assessment: a critical review, and a descriptive model towards a sustainable landscape vision (three-dimensional agrivoltaic patterns). Sustainability 13:6871
8. Abidin MAZ, Mahyuddin MN, Zainuri MAAM (2021) Solar photovoltaic architecture and agronomic management in agrivoltaic system: a review. Sustainability 13:7846
9. Walston LJ, Barley T, Bhandari I et al (2022) Opportunities for agrivoltaic systems to achieve synergistic food-energy-environmental needs and address sustainability goals. Front Sustain Food Syst 6:374
10. Trommsdorff M, Gruber S, Keinath T, et al (2022) Agrivoltaics: Opportunities for agriculture and the energy transition. Freiburg
11. Adeh EH, Good SP, Calaf M, Higgins CW (2019) Solar PV power potential is greatest over croplands. Sci Rep 9:11442
12. Cramer W, Guiot J, Marini K (2020) Climate and environmental change in the mediterranean basin: current situation and risks for the future. Marseille
13. Solargis (2023) Global solar atlas. https://globalsolaratlas.info/map?c=17.476432,30.585938,2&m=site. Accessed 9 Apr 2023
14. Kumar M, Haillot D, Gibout S (2022) Survey and evaluation of solar technologies for agricultural greenhouse application. Sol Energy 232:18–34
15. Emmott CJM, Röhr JA, Campoy-Quiles M et al (2015) Organic photovoltaic greenhouses: a unique application for semi-transparent PV? Energy Environ Sci 8:1317–1328
16. Marrou H, Guilioni L, Dufour L et al (2013) Microclimate under agrivoltaic systems: is crop growth rate affected in the partial shade of solar panels? Agric For Meteorol 177:117–132
17. Weselek A, Bauerle A, Hartung J et al (2021) Agrivoltaic system impacts on microclimate and yield of different crops within an organic crop rotation in a temperate climate. Agron Sustain Dev 41:59
18. Trommsdorff M, Kang J, Reise C et al (2021) Combining food and energy production: design of an agrivoltaic system applied in arable and vegetable farming in Germany. Renew Sustain Energy Rev 140:110694
19. Ferrara G, Boselli M, Palasciano M, Mazzeo A (2023) Effect of shading determined by photovoltaic panels installed above the vines on the performance of cv. Corvina (Vitis vinifera L.). Sci Hortic (Amsterdam) 308:111595
20. Hudelson T, Lieth JH (2021) Crop production in partial shade of solar photovoltaic panels on trackers. AIP Conf Proc 2361:080001
21. Jo H, Asekova S, Bayat MA et al (2022) Comparison of yield and yield components of several crops grown under agro-photovoltaic system in Korea. Agriculture 12:619
22. Jiang S, Tang D, Zhao L et al (2022) Effects of different photovoltaic shading levels on kiwifruit growth, yield and water productivity under "agrivoltaic" system in Southwest China. Agric Water Manag 269:107675
23. Trypanagnostopoulos G, Kavga A, Souliotis M, Tripanagnostopoulos Y (2017) Greenhouse performance results for roof installed photovoltaics. Renew Energy 111:724–731

24. Cossu M, Yano A, Solinas S et al (2020) Agricultural sustainability estimation of the European photovoltaic greenhouses. Eur J Agron 118:126074
25. López-Díaz G, Carreño-Ortega A, Fatnassi H et al (2020) The effect of different levels of shading in a photovoltaic greenhouse with a north–south orientation. Appl Sci 10:882
26. Ezzaeri K, Fatnassi H, Wifaya A et al (2020) Performance of photovoltaic canarian greenhouse: a comparison study between summer and winter seasons. Sol Energy 198:275–282
27. Tang Y, Li M, Ma X (2019) Study on photovoltaic modules on greenhouse roof for energy and strawberry production. E3S Web Conf 118:03049
28. Alinejad T, Yaghoubi M, Vadiee A (2020) Thermo-environomic assessment of an integrated greenhouse with an adjustable solar photovoltaic blind system. Renew Energy 156:1–13
29. Horowitz K, Ramasamy V, Macknick J, Margolis R (2020) Capital costs for dual-use photovoltaic installations: 2020 benchmark for ground-mounted PV systems with pollinator-friendly vegetation, grazing, and crops. Denver
30. Bellini E (2021) Cost comparison between agrivoltaics and ground-mounted PV. In: PV Mag. https://www.pv-magazine.com/2021/03/26/cost-comparison-between-agrivoltaics-and-ground-mounted-pv/. Accessed 8 Apr 2023
31. Scharf J, Grieb M, Fritz M (2021) Agri-photovoltaik—stand und offene fragen. Straubing
32. Li C, Wang H, Miao H, Ye B (2017) The economic and social performance of integrated photovoltaic and agricultural greenhouses systems: case study in China. Appl Energy 190:204–212
33. International Renewable Energy Agency (IRENA) (2022) World energy transitions outlook 2022: 1.5°C pathway. Abu Dhabi
34. Precedence Research (2022) Solar photovoltaic (PV) market—global industry analysis, size, share, growth, trends, regional outlook, and forecast 2022–2030. Ottawa
35. REN21 (2023) Renewables 2023 global status report
36. U.S. Department of Enegy (2022) Market research study: agrivoltaics
37. Precedence Research (2022) Agrivoltaics market - Global industry analysis, size, share, growth, trends, regional outlook, and forecast 2022–2030. Ottawa
38. Precedence Research (2023) Agrivoltaics market is expected to increase at a 12.15% of CAGR by 2030. https://www.precedenceresearch.com/press-release/agrivoltaics-market. Accessed 15 Jul 2023
39. Precedence Research (2022) Commercial greenhouse market—global industry analysis, size, share, growth, trends, regional outlook, and forecast 2022–2030. https://www.precedenceresearch.com/commercial-greenhouse-market. Accessed 15 Jul 2023
40. Parvatha Reddy P (2016) Sustainable crop protection under protected cultivation. Springer
41. freepik.com Couple of workers standing in garden near flowers and talking. https://www.freepik.com/free-photo/couple-workers-standing-garden-near-flowers-talking_7286290.htm#from_view=detail_alsolike. Accessed 9 Apr 2023
42. Pascaris AS, Schelly C, Burnham L, Pearce JM (2021) Integrating solar energy with agriculture: industry perspectives on the market, community, and socio-political dimensions of agrivoltaics. Energy Res Soc Sci 75:102023
43. Ulucak ZS, Yucel AG (2021) Can renewable energy be used as an effective tool in the decarbonization of the Mediterranean region: fresh evidence under cross-sectional dependence. Environ Sci Pollut Res 28:52082–52092
44. Observatoire Méditerranéen de l'Energie (2021) Mediterranean energy perspectives to 2050 (2021 edition). Paris
45. ECMWF & the Union for the Mediterranean (2022) Paving the way for a renewable energy transition in the Mediterranean. https://stories.ecmwf.int/paving-the-way-for-a-renewable-energy-transition-in-the-mediterranean/index.html. Accessed 9 Apr 2023
46. Wikipedia (2023) Solar power by country. https://en.wikipedia.org/wiki/Solar_power_by_country. Accessed 9 Apr 2023
47. Mrabet R, Savé R, Toreti A, et al (2020) Climate and environmental change in the Mediterranean Basin–Current situation and risks for the future. Marseille

48. The world bank (2020) Agricultural land (% of land area). https://data.worldbank.org/indicator/AG.LND.AGRI.ZS?type=shaded&view=map&year=2020. Accessed 9 Apr 2023
49. GRID-Arendal (2013) Agriculture and population in the Mediterranean basin. https://www.grida.no/resources/5906. Accessed 9 Apr 2023
50. Antonelli M, Basile L, Gagliardi F, Isernia P (2022) The future of the Mediterranean agri-food systems: trends and perspectives from a Delphi survey. Land Use Policy 120:106263
51. Booth A (2021) The reason we're running out of farmers. https://www.bbc.com/future/bespoke/follow-the-food/the-reason-we-are-running-out-of-farmers/#:~:text=Worldwide%2C the percentage of people,to work on farms anymore. Accessed 9 Apr 2023

Chapter 4
An Overview of Solar Cell Technologies Toward the Next-Generation Agrivoltaics

This chapter accounts for an overview of solar cell technologies developed up to now, from the mature silicon-based solar cells to the third-generation photovoltaics. Herein, the most important characteristics, current challenges and strategies for further improvement of each solar cell technology are discussed, aiming to demonstrate possible manufacturing options for the development of the next-generation agrivoltaics. The short life-cycle assessment review on solar cell technologies that is also included at the end of this chapter, provides the reader with a more comprehensive point of view toward the development of greener agrivoltaic energy systems.

4.1 Introduction

It has been 184 years since Alexandre Edmond Becquerel first observed the photovoltaic (PV) effect in 1839 by immersing a system of electrodes in a conductive solution and exposing them to light [1]. Nevertheless, the story of practical solar cell devices goes back to 1954, when Bell Laboratories demonstrated the first silicon solar cell [2]. This was the time when the PV field was born, empowering humanity to realize the dream of using solar energy to convert it to electricity for daily life. Since then, various solar cell technologies have been developed with the aim of producing electric power from the sunlight more efficiently, in a reliable manner and at a lower cost. Nowadays, the world is going through a euphoric rush to install PV devices almost everywhere, from rooftops of houses to large glass buildings, parking spaces, vehicles, overwater bodies and many other applications, aiming at sustainable development.

The evolution of solar PVs through the past almost seventy years has demonstrated a wide variety of technologies, each of them providing unique technical attributes. Up to now, solar cell devices have been categorized into various groups depending on the considered classification criteria. Figure 4.1 illustrates the classification of solar

D. A. Chalkias and E. Stathatos, *The Emergence of Agrivoltaics*,
Green Energy and Technology, https://doi.org/10.1007/978-3-031-48861-0_4

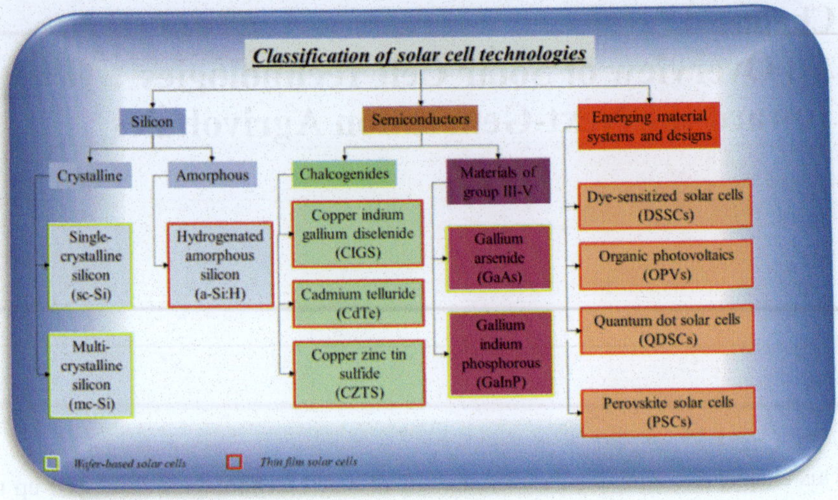

Fig. 4.1 Classification of solar cell technologies according to the main active material employed

cell devices depending on the key materials employed. The current certified record power conversion efficiencies (PCEs) for the most important solar cell technologies are presented in Fig. 4.2.

Today, efficiency and cost are the two main attributes that are widely used to classify solar cell technologies. Thus, three different PV generations are established. The first generation of PVs regards wafer-based solar cells and considers systems characterized by high costs but also high efficiencies. The second generation of PVs regards thin-film solar cells, which are characterized by lower prices, but their efficiency is moderate. Finally, the third PV generation focuses on solar cell technologies developed for attaining high efficiency at low manufacturing costs. This category includes revolutionary thin-film devices, as well as a range of exotic technologies, such as hot-carrier collection, carrier multiplication, spectral splitting and thermophotovoltaics, each of them giving new insights for an even more competitive renewable energy production. These three PV generations are frequently shown as shaded areas on the efficiency versus cost plot. Figure 4.3 depicts these areas as they were originally described prior to 2000. As efficiencies increase and prices decrease over time, all technologies tend to move toward the upper-left corner. Today, due to the market dynamics, the average commercial module costs for both first and second PV generations tend to cluster along a single $ per W line. On the other hand, no third-generation technology has yet entered the "III" zone [4].

An alternative method to classify solar cell technologies is according to the complexity of the employed materials, i.e., the number of atoms in a single cell, molecule, or another repeating unit, as shown in Fig. 4.4. According to this model, the complexity of solar cell technologies ranges from elemental (lowest) to nanomaterial (highest). At this point, it is important to mention that processing complexity is

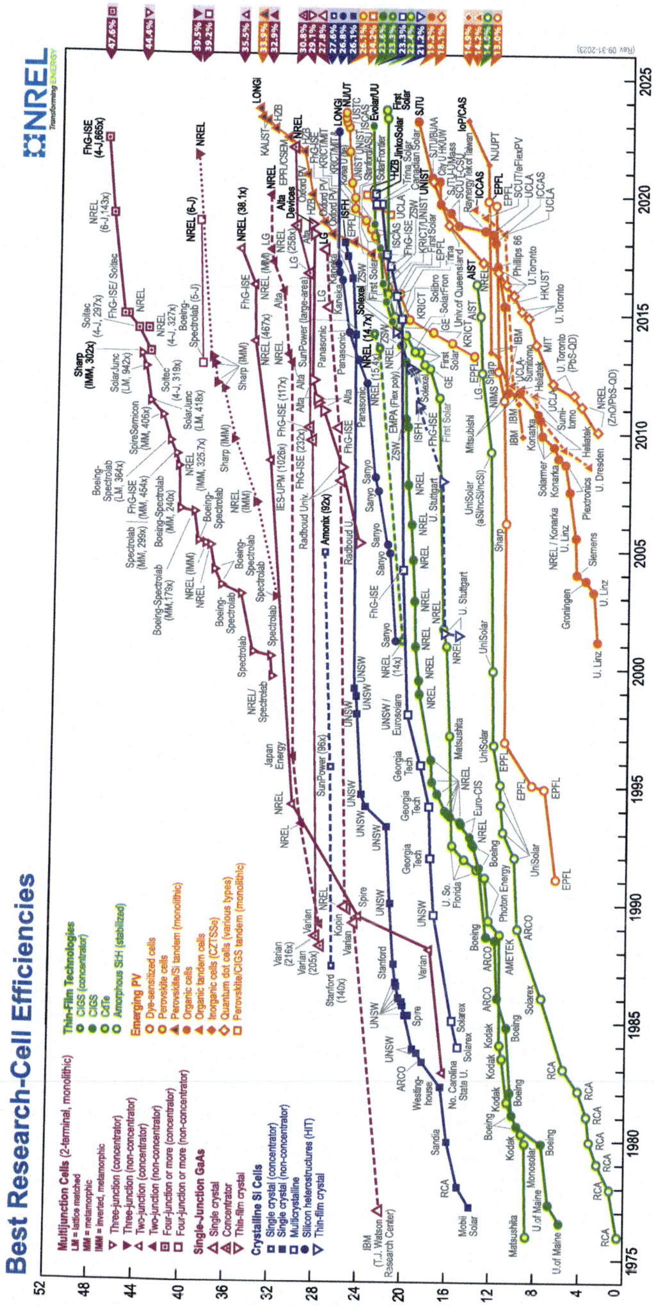

Fig. 4.2 Certified efficiency records for the different solar cell technologies [3]

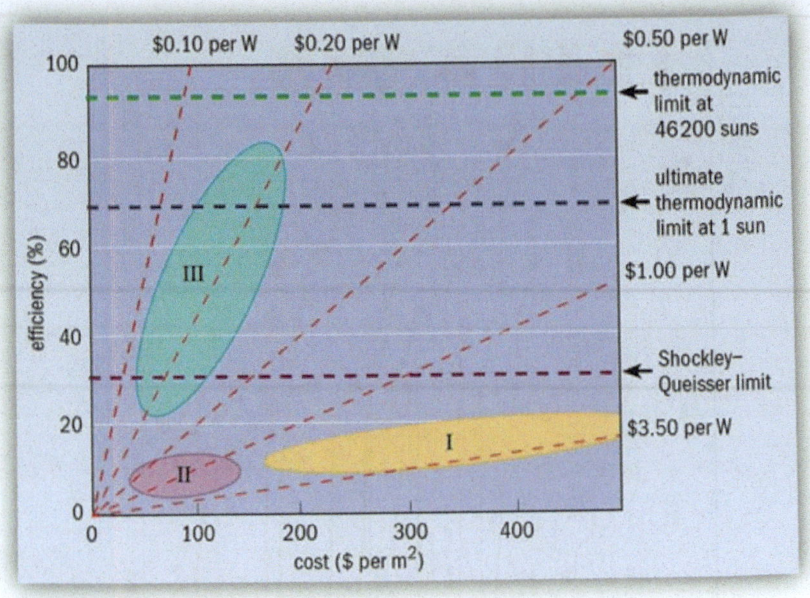

Fig. 4.3 Efficiency versus cost chart for the different generations of solar cells [5]

different from material complexity. Higher material complexity is not always "better". Solar cell efficiency and technological maturity sometimes vary in an inverse relationship to complexity. However, increased material complexity can provide a number of unique and valuable characteristics, improved defect tolerance, reduced material consumption, visible transparency, etc. [4].

4.2 Crystalline Silicon Solar Cells

Since the early years of development of the PV field, crystalline silicon (c-Si) solar cells have been considered the workhorse of the PV industry and will remain the technology leader until a more efficient and cost-effective alternative is developed [6]. Today, c-Si solar cells have overshadowed the global PV market, which now relies on about 90% on silicon. The dominance of c-Si PVs has historical reasons, such as the early development of this type of technology and the continuous progress in the field, as well as the concomitant rise of the microelectronics sector [7]. Compared to other existing PV materials, c-Si is among the most suitable candidates for solar cells fabrication due to its natural abundance (about 25% of Earth's crust), high stability, non-toxicity and cost competitiveness [7]. Its indirect bandgap of 1.12 eV fits well with the solar spectrum, providing the capability to capture a wide part of sunlight that

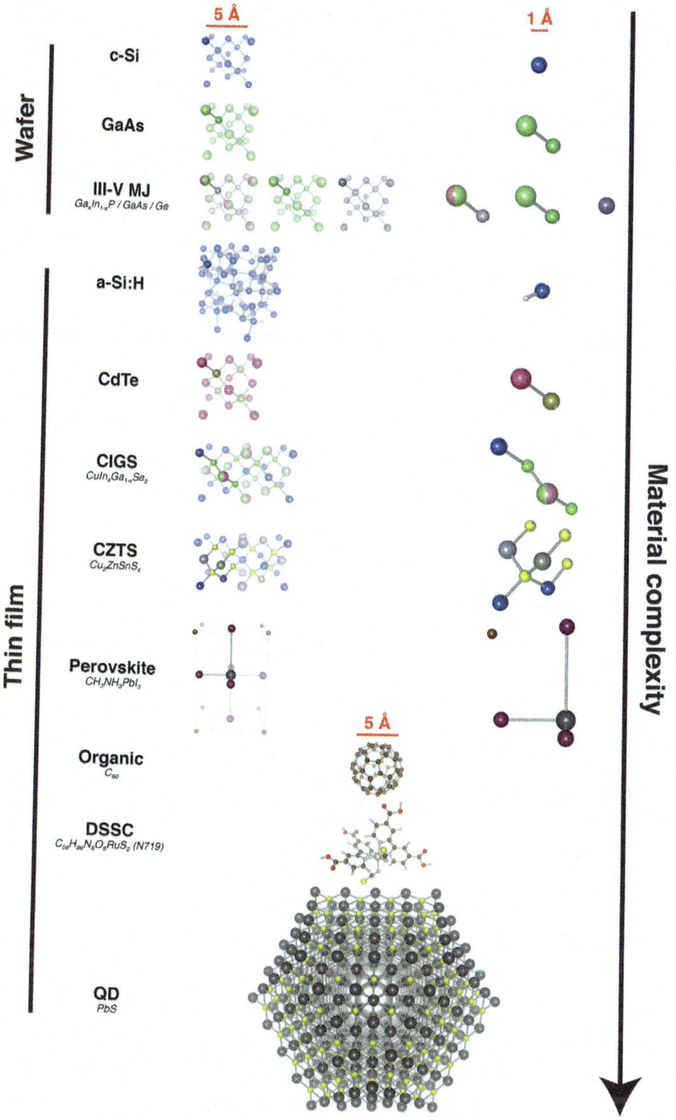

Fig. 4.4 Classification scheme for solar cell technologies according to the complexity of the materials employed [4]

can benefit the production of high photo-current. Besides, with warranties spanning 20–30 years, these devices also provide an exceptional level of reliability, outperforming even most of the highest robust products available in the market, regardless of the application field [8]. Some of the key disadvantages of the under-consideration technology are the rigid, brittle and fairly thick design of these solar cells; the latter is needed to capture the incident light effectively. These characteristics culminate in a high capital outlay, low power-to-weight ratios and restrictions in their design, including the ability to develop flexible devices. Besides, some of the primary technical challenges of c-Si solar cells are also the stringent material purity requirements, combined with the high material consumption, batch-based manufacturing, limitations on module form factor and relatively low throughput production [4].

The c-Si PVs are subdivided into cells made of multi-crystalline silicon (mc-Si) and single-crystalline silicon (sc-Si). For decades, mc-Si dominated the market. However, sc-Si has become the predominant PV material in recent years [7]. Today, the methods for c-Si solar cell and module manufacturing are highly developed and reliable. Typically, sc-Si solar cells are fabricated by adapting the Czochralski, float-zone or Bridgman techniques [9]. They are endowed with greater crystal quality than mc-Si solar cells, which boosts charge extraction and PCEs, but necessitates wafers that are 20–30% more costly than mc-Si ones [4]. As sc-Si PVs are composed of single silicon crystals, their production is more complex and expensive. Currently, the efficiency record for sc-Si solar cells (non-concentrator) is 26.1% [3]. Besides their better efficiency, further benefits of sc-Si solar cell technology over mc-Si one include their higher longevity and greater heat resistance. Regarding mc-Si solar cells, they are comprised of a number of randomly oriented grains or crystals, which introduce barriers that impede charge extraction and electron flow, stimulating them to recombine with holes and diminishing the solar cell power output [4]. This renders mc-Si solar cells less efficient in direct sunlight, but, at the same time, cheaper. Moreover, their performance in low-light conditions is often superior to that of sc-Si devices. Today, the record efficiency of mc-Si solar cells is 23.3% [3].

The c-Si solar cells can also be subdivided according to their design to n-type and p-type. An n-type c-Si solar cell has a silicon base usually infused with phosphorus, with the top silicon layer of the wafer being infused with boron to develop the p–n junction. On the other hand, p-type c-Si solar cells are developed using a silicon base infused with boron atoms, while the top silicon layer is infused with phosphorus for the p–n junction formation [10]. Although the first solar cell that was invented in Bell Laboratories in 1954 was of n-type design, the p-type structure became predominant the next years, and up-to-now has the highest market share [11]. This is because p-type cells were early found to be more robust to space radiation. Thus, more investments were available in this technology, which resulted in a high amount of data and information about their performance and stability. Today, more and more solar cell developers are adopting the n-type design because of some important benefits. Of the main advantages of n-type solar cells is the almost absent boron-oxygen defects. In contrast, p-type cells present a significant concentration of boron-oxygen defects. Under light exposure, these defects are increased at an important rate, which subsequently results to a further decrease of the performance of the devices. Instead, n-type solar cells use mainly phosphorus instead of boron for silicon doping, which ensures a kind of immunization to boron-oxygen defects. Thus, this technology is considered more resistant to light-induced degradation [11]. Furthermore, p-type solar cells are more prone to metallic impurities, such as interstitial iron, which can reduce their lifetime [10]. Nevertheless, the main disadvantage of n-type solar cells is their cost. According to the International Technology Roadmap for Photovoltaic (ITRPV), p-type sc-Si will dominate over 30% of the market through 2028, while n-type c-Si will rise to around 28% from barely 5% in 2017. This conforms to the industry demands for higher efficiency modules, which favors n-type designs entering the mainstream [11].

In the race for developing high-efficiency silicon PVs, heterojunction (HJ) solar cell design was early employed as an upgrade to the homojunction c-Si solar cells. Silicon HJ PVs are developed by combining two different technologies (i.e., two different semiconductor materials). More specifically, they are made of c-Si wafers using passivated contacts of thin-film silicon alloys, such as hydrogenated amorphous silicon (a-Si:H), nano-crystalline silicon (nc-Si:H) or silicon oxide (nc-SiO_x:H) [12]. Due to the high quality of the c-Si wafer and the excellent surface passivation, silicon HJ solar cells provide a series of important features. Some of them are the high PCEs, with the photo-voltages exceeding 740 mV, the better thermal stability, with the temperature coefficients to be determined at low values (<0.3%/°C), and the lower production costs compared to conventional Si-based devices [12].

The aim of the current research on c-Si solar cells is to develop devices at lower manufacturing costs and with less complexity, to improve the modular PCEs, to use less silicon per watt, as well as to rely less on silver for contact metallization. Novel technologies are continuously introduced towards the fabrication of higher efficiency c-Si solar cells that can reach the theoretical PCE limit, which is about 30% (see Fig. 4.5) [14]. Aluminium back surface field (Al-BSF) cells, passivated emitter rear contact (PERC) cells, tunnel oxide passivated contact (TOPCon) cells, interdigitated back contact (IBC) cells, heterojunction with intrinsic thin-layer (HIT) cells and heterojunction cells with interdigitated back contacts (HBC) are some of the technologies that have been developed over the last decades toward the fabrication of high-efficiency c-Si solar cells [13].

For many years, Al-BSF technology was employed to achieve high-efficiency c-Si solar cells. The Al-BSF is one of the most common solar cell technologies, given its relatively simple manufacturing, which can be applied to realize competitive performance in solar cells. The main driver for the enormous success of this solar cell technology is also the structure and process tolerance against fluctuations in wafer quality, such as variations in the concentration of base material doping. Nevertheless, critical is the availability of the accompanying manufacturing. None of the vital structure elements or process sequences is severely protected by patents or other legal issues. This enables many equipment and material manufacturers to compete for the finest and least expensive products. Due to the enormous demand for production technology on the market, these factors were critical to a rapid rise in production capacity. Today, the typical efficiencies for this cell structure are about 20% [7].

Due to the prospects of attaining higher efficiency, a switch from Al-BSF toward the PERC technology started to take place when reliable and low-cost production was available. Solar cell efficiencies exceeding 23% were already achieved by PERC technology from the early 1990s [16]. The introduction of a back surface that is dielectrically passivated and has local point contacts set the backbone of this concept. In comparison to cells without dielectrically passivated surfaces, cell characteristics

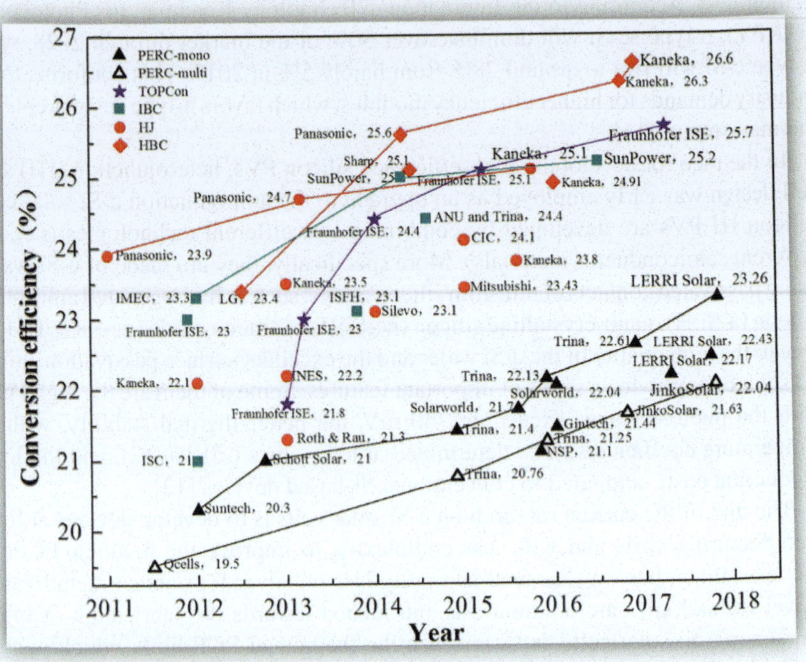

Fig. 4.5 Noteworthy efficiency records for crystalline silicon solar cells using different technologies. The lines serve as visual aids [13]

were improved because of a decreased surface recombination velocity and enhanced optical performance [7]. Those layers improve the electrons movement in the device, as well as bounce light back into the cell, giving it a second chance to capture electrons that would otherwise simply pass through. As an upgrade to PERC technology, passivated emitter rear locally diffused (PERL) cells and passivated emitter rear totally diffused (PERT) cells were developed. PERL technology cells were introduced as an upgrade of PERC ones to reduce further the point contacts recombination and to allow for the usage of higher resistivity materials in the solar cell design. Nowadays, solar cell efficiencies of up to 25% have been attained using this technology [7]. On the other side, in the PERT structure, the rear surface is "totally diffused" with either boron (p-type) or phosphorus (n-type). PERT technology is often applied to n-type c-Si solar cells. This is because n-type silicon wafers provide some important characteristics over p-type silicon wafers, such as a better tolerance to metallic impurities, lower temperature coefficient and lower light-induced degradation. Nevertheless, the production of PERT technology cells is of higher cost compared to the PERC since additional novel processes are needed during manufacturing [17]. Simultaneously, additional novel concepts are continuously developed on the basis of PERC technology. Of the most important ones are bifacial PERC solar cells that are increasingly employed nowadays in the fabrication of high-efficiency c-Si solar cells. Bifaciality

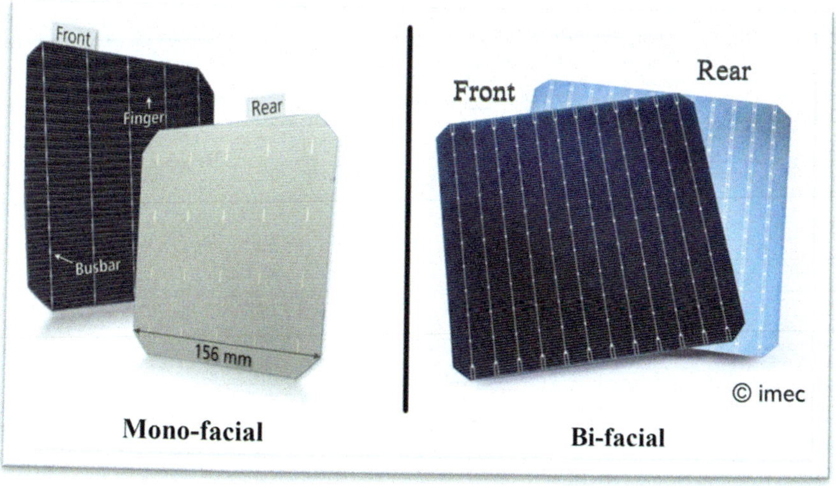

Fig. 4.6 Images of the front and rear side of a mono-facial and a bi-facial crystalline silicon solar cell [7, 15]

can be accomplished in a PERC structure by substituting the full-area Al rear contact with an Al grid, providing a rear side efficiency usually between 70 and 80% (see Fig. 4.6) [7]. With the turn of the millennium, pathways toward the industrialization of bifacial solar cells, including bifacial PERC cells and modules, started to be laid and able to meet international commercialization standards.

The past years of development of c-Si solar cell technology have shown that just when the performance seems to have reached a peak, the researchers in the field provide new pathways to attain higher performance at lower manufacturing costs. TOPCon technology is considered the next generation of PERC technology. The basic architecture of TOPCon was introduced in 2013 [18]. TOPCON technology is basically the next generation of PERC, and like its forbear, it can be applied to cells manufactured in the traditional way. Because TOPCon is the next step after PERC, it does not add a great deal of additional cost to the finished product. The easiest way to understand the TOPCon technology is to compare it to the PERT. The introduction of an ultrathin SiO_2 layer and a doped poly-crystalline-silicon (poly-Si) layer is the difference compared to PERT technology (see Fig. 4.7). Poly-Si is used as a current collector since it is highly conductive due to its high doping. In an n-type TOPCon cell, the poly-Si layer is doped with phosphorous to achieve both a back surface field and surface passivation. A significant gain in solar cell efficiency can be achieved by TOPCon technology compared to the previously developed technologies, which can be determined to a higher than 1% in absolute values [7].

IBC technology is of the most important ways to realize high-efficiency c-Si solar cells. In an IBC solar cell architecture, both negative and positive metal contacts are moved to the back of the solar cell side, so that the front side completely faces the light. This not only results in more efficient solar cells since the active-to-aperture

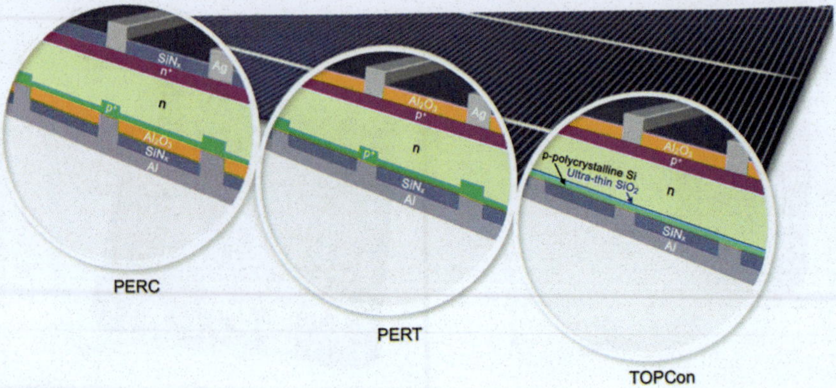

Fig. 4.7 Schematic cross-sectional representation of different crystalline silicon solar cell technologies [7]

area ratio of the devices is increased, but also in improved aesthetics (no visible contacts) (see Fig. 4.8). The main component employed in an IBC c-Si solar cell is an n-type wafer with boron-diffused emitters, while there is also an antireflection layer on the front side of the solar cell to improve the light-trapping effect. Additionally, the front and back thermal SiO_2 passivation eliminates surface recombination and improves the long wavelength response. Compared to the conventional c-Si solar cell architecture, in IBC c-Si solar cells, the recombination rate on the front surface has a greater effect on the performance of the devices. This is because the front surface is far away from the p–n junction, which is located on the rear side. For this reason, a better surface passivation strategy is required to prevent front surface recombination, while the integration of high-quality materials can provide long diffusion lengths, so that the photo-generated minority carriers have enough time to reach back contacts before recombination [13].

Regarding the silicon HJ solar cells, heterojunction with intrinsic thin-layer (HIT) was of the first developed designs in this group of solar cells, in the early 1990s decade. These solar cells are comprised of a thin c-Si wafer surrounded by ultra-thin intrinsic a-Si layers and doped (n-type and p-type) a-Si layers [13]. The a-Si layers are formed on both sides of the wafer using low-temperature processing. In this way, the degradation of the crystalline bulk material is avoided. HIT solar cells provide multiple advantages over the standard ones [13]. The most important is that they can provide higher PCEs than the corresponding ones of homojunction c-Si devices. This is because the bandgap of a-Si is in the range of about 1.55–1.87 eV. In this way, the light absorption by the device in the visible spectrum, where the intensity of terrestrial solar radiation is high, is increased. Using the a-Si layers, they also provide a significant decrease in surface and interface recombination, which results in the production of higher photo-voltages. It is important that the aforementioned can be achieved without increasing the manufacturing costs at significant levels, setting these devices as a high-efficiency and low-cost alternative to the conventional

Fig. 4.8 Image of interdigitated back contact (left) and passivated emitter rear contact (right) crystalline silicon solar cells [19]

homojunction c-Si devices. Nevertheless, stability issues related to the degradation of a-Si usually appear during their lifetime, which raises questions about their reliability.

Given that it combines the benefits of both the BC and HJ structures, HBC solar cell architecture is one of the predominant candidates for developing high-efficiency c-Si solar cells today. This is attributed to the ability of high photo-current production by the employment of the BC design, due to the absence of front-side electrodes that decrease the active area of the solar cell. At the same time, the HJ structure can lead to the production of high photo-voltage due to the superior passivation compared to conventional devices [20]. By applying HBC architecture, c-Si solar cells can now attain efficiencies higher than 26.6%, which is a record value for this technology today [21].

Figure 4.9 presents the current and forecast market shares for different silicon solar cell technologies according to the ITRPV report for 2022. As is demonstrated, the conventional BSF design is about to phase out from the PV market within 2022. In its place, the now matured concept of diffused and passivated p–n junction designs will be further used in the mainstream, with the PERC/PERL/PERT/TOPCon cell technologies to have more than 85% market share. On the other hand, HJ technology is expected to reach a market share of about 19% soon, while other high-efficiency technologies, such as BC technology, will follow with about 5% each.

Table 4.1 summarizes the record announcements and the average solar cell and module efficiencies available on the PV market for the different silicon solar cell designs.

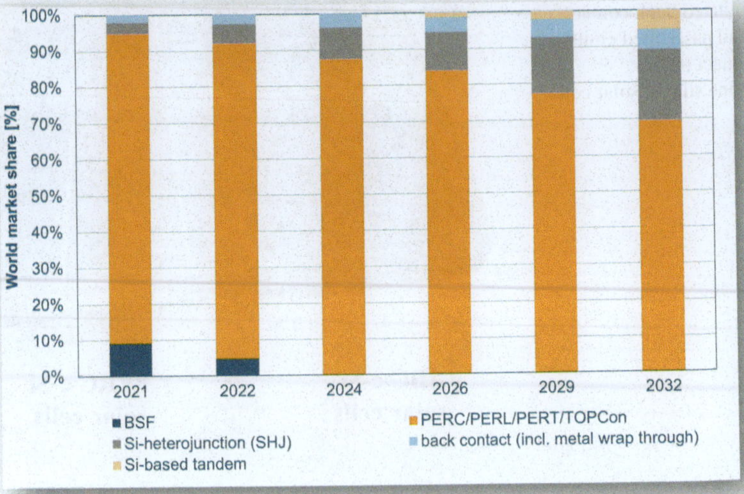

Fig. 4.9 Current and forecast market share of different solar cell technologies [22]. *Source* Adapted from the 13th edition ITRPV, 2022

Table 4.1 Efficiency of silicon solar cells and modules based on different cell designs [23]

Si solar cell design	Record cell efficiency (%)	Average cell efficiency (%)	Available module efficiency (%)
Al-BSF	20.3	20	<20
PERC	24.1	23	20.7
TOPCon	25.7	24	21.4
IBC	25.2	24	21.3
HJ	25.3	24	21.8
HBC	26.6	25	22.4

4.3 Commercialized Thin-Film Solar Cells

The c-Si solar cells nowadays account for a vast part of the global PV market. However, new technologies are continuously developed that open up the way for lower-cost PV energy production. Among the PV technologies, thin-film designs are the most promising options for lowering the costs of light-to-electricity conversion [24]. With the aforementioned potential, along with the prospect of attaining a high efficiency, solar cells based on thin semiconducting films in the last years have occupied about 10% of the global solar modules production capacity [25]. Hydrogenated amorphous silicon (a-Si:H), copper indium gallium diselenide (CIGS) and cadmium telluride (CdTe) are the top commercial thin-film solar cell technologies of today (see Table 4.2). Due to their superior light absorption ability (10 to 100 times higher) compared to the c-Si, these materials can be used in thin-film solar cell designs, with thicknesses of only a few microns. Namely, the incident light can be absorbed almost completely in a film fabricated by the aforementioned materials, with the device thickness to be at the level of 1 μm, while c-Si solar cells need thicknesses of around 50 μm, and for handling reasons, 200 μm. These thin-film technologies have a key advantage over the c-Si wafer design since they use fewer raw materials and require less sophisticated manufacturing processes. Thin-film modules can now be fabricated in a highly automated and streamlined manner, resulting in low-cost-per-watt modules. Compared to c-Si counterparts, their handling is also easier, more flexible and less susceptible to damage [26]. They can also benefit from their lightweight and the capability to develop flexible devices, which open the way for new applications. Nevertheless, due to their lower PCE compared to c-Si solar cells, they have a higher area-dependent balance of system (BOS) cost. Besides, their sensitivity to oxygen and moisture appears as another drawback, making hermetic encapsulation imperative to ensure long-term stability, which subsequently increases the manufacturing cost. Their reliance on rare elements, such as indium and tellurion, and toxic elements, like cadmium, also limit their widespread production. Current innovation and development opportunities in thin-film technologies include improvements in both materials and cells architecture. Simultaneously, the development of new materials with the same processing ease as existing ones is pertinent in order to reduce the reliance on rare elements [26].

Throughout the history, the a-Si has the longest time in the commercial arena, starting with its application in watches and calculators in the late 1980s. Thin-film solar cells based on a-Si:H has been used for decades, using a-Si:H both as a doped material and as an intrinsic absorbent layer. Whereas for many years, their efficiency was boosted by the development of higher-quality absorbent layers, current advances are attributed to a better understanding of the device interfaces, allowing for more precise engineering [27]. Highly efficient, lightweight and flexible solar cells can now be developed on the basis of thin-film a-Si:H layers of about 300 nm thickness, which can absorb almost 90% of above-bandgap photons in a single pass [4]. Moreover, a-Si:H solar cells can be combined with nano-crystalline silicon or amorphous silicon–germanium alloys to form multi-junction solar cells, without the need for

Table 4.2 Comparison of hydrogenated amorphous silicon, copper indium gallium diselenide and cadmium telluride properties

	a-Si:H	CIGS	CdTe
Bandgap (eV)	≈1.7 (direct)	1.0–1.7 (direct)	≈1.5 (direct)
Appropriate thickness (μm)	1	1–2	3–5
Temperature coefficient (%/°C) (efficiency, from 25 to 65 °C)	−0.13	−0.36	−0.21
Toxicity	None	None	Cadmium

lattice matching. Today, the record efficiency for a-Si:H solar cells exceeds 14%, while their global market share is about 1% [3, 28].

The success of a-Si:H thin-film materials relies mainly on the ability to apply high throughput processes and large-area manufacturing. Whereas c-Si solar cells are typically developed from self-supporting slices sawn from silicon ingots, the silicon layers in thin-film silicon PVs are deposited from the gas phase, using mainly SiH_4 feedstock gas with H_2 dilution, under much lower-temperature processing [29]. The chemical vapor deposition (CVD) methods, particularly the very high-frequency plasma-enhanced CVD, enable their simple scaling up, where the moveable substrate of large size and long depositions without downtime is feasible. Particularly interesting is a-Si:H thin-film deposition on flexible substrates, from papers (e.g., cellulose papers) and temperature-sensitive plastics (e.g., polyethylene terephthalate (PET) and polyethylene naphthalate (PEN)), to high temperature-processable plastics (e.g., polyimide and Kapton). Despite the fact that a large number of different solar cell material alternatives have been developed, a-Si:H solar cells have shown products on a quite larger scale, e.g., a-Si:H solar modules of higher than 5 m^2 size on both rigid and flexible substrates [30].

Thin-film solar cells based on a-Si:H are developed as PIN junction [31]. The PIN junction consists of three differently doped regions, namely an intrinsic or undoped layer sandwiched between a p- and an n-doped material. The intrinsic layer is necessary to allow the separation and collection of the photo-generated carriers. Depending on the growth of the thin films on the top of each other, two configurations exist, with no fundamental difference in their working principle. These are labeled as "p-i-n" or "superstrate" and "n-i-p" or "substrate", where p = p-doped layer, i = intrinsic layer and n = n-doped layer (see Fig. 4.10). The thickness of the layers is approximately: substrate ≈ 0.25–3 mm, TCO ≈ 800 nm, p ≈ 10 nm, i ≈ 300 nm, n ≈ 20 nm and Ag ≈ 800 nm [31]. In contrast to c-Si solar cells, the diffusion lengths in a-Si:H are quite short and even shorter in doped a-Si:H. This is because the material is characterized by a significant number of mid-gap states (defects). Charge carriers generated by illumination are driven and separated by a high electric field developed across the i-layer when it comes in contact with the p- and n-doped layer. This field makes the photo-generated carriers drift toward the contacts. Thus, a-Si:H are called drift cells or field cells since the generated minority carriers flow as pure field current, unlike the c-Si solar cells, where the minority carriers reach the space charge region by diffusion and then are separated by the electric field (diffusion cells) [31].

Today, almost all a-Si:H solar cells are illuminated through the p-layer. This is due to a combination of two reasons [31]. Firstly, the front-side illuminated layer is that it absorbs most of the incident light. Secondly, the mobility of holes is lower than for the electrons in a-Si:H. Consequently, to get the photo-generated charges collected efficiently, the average distance for holes to travel should be shorter. Namely, most charges should be generated close to the p-layer, which is achieved by device illumination from the p-layer. A p-i-n device is usually prepared on a transparent conductive oxide (TCO) glass substrate, which fulfills a couple of requirements. These are the high transparency, adequate conductivity and suited surface texture to provide effective light trapping by scattering the incident light and thereby increasing the optical

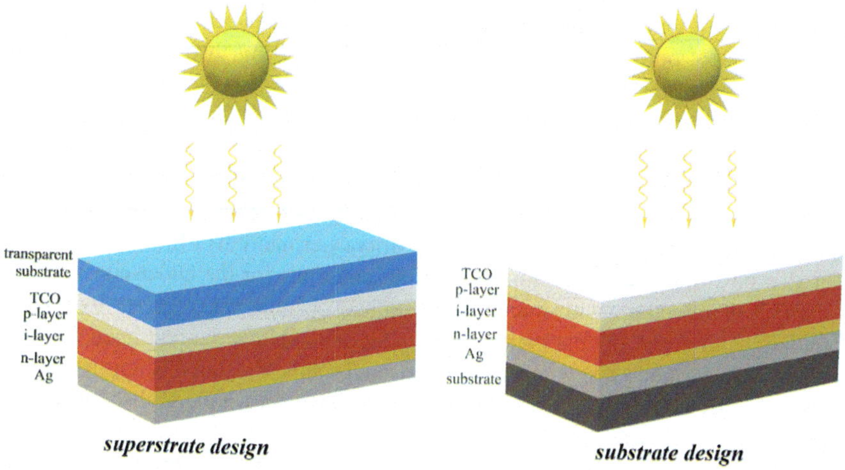

Fig. 4.10 p-i-n (superstrate) and n-i-p (substrate) design of amorphous silicon solar cells, where, in both cases, the device is illuminated from the p-side

path length. In contrast, n-i-p devices can be developed on a variety of lower-cost substrates, such as glass, aluminium, plastics, etc. Regarding the p-layer and n-layer, the electron–hole pairs generated therein do almost not contribute to the photo-current production since most of them are recombined. Hence, these layers should be ultra-thin and exhibit a wide bandgap in order to avoid optical losses. On the other hand, the thickness of these layers should be adequate to build up an efficient electric field across the i-layer. Regarding the i-layer, it should demonstrate high charge carrier mobility and a long lifetime in order to provide an efficient collection of generated carriers. To attain as much current as possible, the absorption coefficient combined with the i-layer thickness should be tuned to allow high photo-current production without deteriorating the fill factor [31].

Irrespective of the process used for the deposition of Si-based thin films, dilution of the Si-containing source gas with H_2 results in thin films with different ordering, morphologies and textures than neat a-Si. By applying moderately H_2-diluted gas mixtures, the resulting material (often referred to as proto-crystalline silicon (pc-Si:H)) has a higher medium-range structural order than the standard a-Si. Besides, its stability against light-induced degradation is also higher than the latter material. At increasing H_2-dilution ratios, a transition from amorphous to crystalline structures is observed. In this transition regime, several material structures can be obtained, such as quasi-amorphous or nano-amorphous silicon (na-Si:H), polymorphous silicon (pm-Si:H), nano-crystalline silicon (nc-Si:H), micro-crystalline silicon (μc-Si:H) and heterogeneous silicon (het-Si:H), depending on whether crystals are present and on the crystalline fraction in the material bulk. Of course, in this way, the characteristics of the resulting material also change [29].

Since the characteristics (including the optical bandgap) of the thin film silicon can be altered by the variation of its crystalline content, this material can easily be employed to build up tandem and multi-junction solar cell architectures. Such cells have great potential for high PCE, mainly because of their spectrum-spitting capability [29]. The deposition of several Si-based thin films on top of each other is found to be an elegant way to overcome the tradeoff between the electrical and optical performance in this type of solar cells. The combination of an a-Si:H top cell with further sub-cells of decreasing bandgaps toward the bottom, as well as the tandem junction of an a-Si:H top cell and a μc-Si:H bottom cell, are widely used approaches [27]. Hydrogenated amorphous silicon germanium (a-Si_xGe_{1-x}:H) alloys can also be employed to develop compatible materials for the under-consideration solar cell architecture, with manipulated optical characteristics; since by varying the composition, the optical band gap of a-Si_xGe_{1-x}:H can be regulated from 1.7 to 1.0 eV. Nevertheless, accompanying drawbacks appear using this approach, such as the serious light-induced instability that usually characterizes these structures compared to the a-Si:H one [32].

On the other hand, CIGS solar cells are a promising and widely used technology for light-to-electricity conversion because of their low-cost, satisfactory efficiency and easy integration into structures. As in the case of a-Si:H, CIGS has a long history of research and commercialization. Their development on glass substrates is now at a mature level, while the industrial work on flexible devices also dates back to the decade of the 1990s. Today, multiple leading manufacturers are working on CIGS solar cell technology. The production cost of these devices can be lower than $0.40/ Watt (excluding BOS and installation charges), while even lower costs with a wider application range can be achieved by their development on flexible substrates [30].

CIGS devices provide many advantages compared to the c-Si technology. At first, they are easy to fabricate by their growing on both rigid and flexible substrates, without the need to apply such strict manufacturing conditions as in the case of c-Si, using a variety of solution and vapor-phase techniques [33]. CIGS semiconductor demonstrates a tunable bandgap of 1.0–1.7 eV and a high absorption coefficient of up to 10^5 cm^{-1}, which sets it an excellent solar cell material for developing tandem devices [34]. CIGS solar cells are also less expensive than c-Si solar cells due to their thin-film design, while their efficiency record is at the same level. Today, the record efficiencies of CIGS solar cells and modules are 23.4% and 19.2%, respectively [3]. It is important that these solar cells exhibit high radiation resistance, which makes them suitable for space applications [4]. Nevertheless, CIGS solar cells are not widely commercialized due to several reasons. Of the main ones is the difficulty of developing high-quality materials on a large scale. The efficiency of commercial large-sized CIGS solar modules is usually much lower than the corresponding of lab-sized devices. A key reason for this efficiency limitation is electronic inhomogeneity [34]. CIGS semiconductor is very sensitive to manufacturing conditions. Thus, considerable research and development efforts are nowadays made in the direction of overcoming the limitations on the chemical and electronic structures of CIGS. Additionally, major technological challenges in CIGS solar cell technology include the difficulty in controlling film stoichiometry and properties, the ease of development of defects, the inadequate understanding of the effects of grain boundaries,

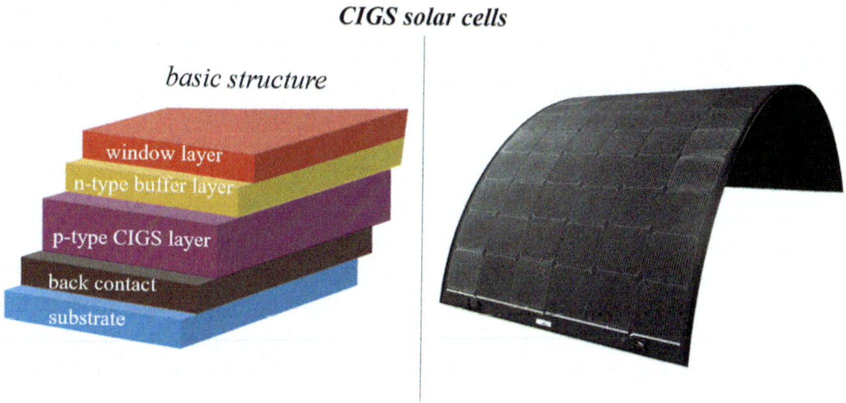

Fig. 4.11 The basic structure of copper indium gallium diselenide solar cells and image of a corresponding solar module fabricated on a stainless steel substrate [35]

as well as the difficulty in developing higher bandgap alloys to give way for the fabrication of multi-junction devices [26].

CIGS solar cells are usually developed via substrate configuration shown in Fig. 4.11, where the incident light comes from the electrode window layer of the device, commonly being a TCO. The light passes through the n-type buffer layer, which is usually an n-type CdS or doped/undoped ZnO or In_2O_3, subsequently reaching the main active layer, i.e., CIGS. By the impact of the electric field produced at the interface CIGS/CdS, electrons are moved from the p-type absorber to the n-type material, and holes in the opposite direction, finally reaching the contacts of the device, whereas back contact is the usually employed molybdenum (Mo) [33]. CIGS can be fabricated by various fabrication techniques. In the group of vacuum-assisted techniques, the most common are physical vapor deposition (PVD), chemical vapor deposition (CVD) and co-evaporation. On the other hand, there is a great effort today to develop CIGS PV devices using non-vacuum techniques. The non-vacuum techniques require much lower energy input and are more compatible with large-scale manufacturing. Usually, non-vacuum techniques follow a two-step process. The first step includes the deposition or printing of the CIGS precursor at low temperatures. This process can be classified as electrodeless deposition, such as electrodeposition and chemical bath deposition, and solution-based deposition, where the most common techniques are spin-coating, doctor-blade coating, spray coating, screen-printing and inkjet-printing [34]. In electrodeless deposition techniques, the precursor material is deposited using complexing agents to decrease all individual precursor element potentials or make them similar to each other. Whereas, in solution-based deposition, the precursor material is initially reduced to a particulate form with the desired stoichiometry. In this way, large-scale production, with low cost and high speed, can be realized. After this first step, the CIGS layer undergoes sintering in a controlled Se atmosphere. Today, under the development of materials science, non-vacuum processes can lead to good quality materials as the corresponding arising from vacuum-assisted processing. This can be achieved by the precise tuning of the compositions and the amount of doping during the synthesis of the CIGS precursor, controlling the Se composition during the selenization process, as well as modifying the surface structure of the CIGS layer by heat treatments [34].

Up until now, many attempts have been made in the direction of tuning the optical and electronic characteristics of CIGS solar cells by structural engineering to attain a more efficient light-to-electricity conversion [34]. At first, this can be achieved by changing the intrinsic structure of CIGS, which can lead to the manipulation of the bandgap of the material. By tuning the Cu/Ga/In composition, a graded bandgap can be obtained. This approach is found to reduce recombination losses and ensure a highly efficient charge collection. Moreover, CIGS bandgap manipulation can be achieved by alkali metal doping, such as Li, K, Na, Cs and Rb. To enhance the performance of the solar cells, manipulation of the surface morphology of the CIGS absorber layer can also be adopted. Controlling Se flux during CIGS growth affects the CIGS film morphology and, subsequently, the solar cell parameters. Nanostructured CIGS film surfaces, such as nanopillar, nanowire and nanotip arrays, enable a shorter carrier diffusion length and can increase solar cell efficiency. Besides, a modification of the elemental concentration at the nanostructured surface can decrease the shunt leakage and series resistance, as well as facilitates the formation of CdCu at the CdS/CIGS interface to enhance carrier extraction, interface passivation and achieve better adhesion. Engineering can also be applied to the structure of the rest materials of the device [34]. The proper selection of the n-buffer layer is determent for the PCE of these solar cells. Even though CdS is considered the best choice as an n-type buffer layer, several Cd-free (toxic-free) materials are tested for the corresponding application, such as $Zn(O, S, OH)_x$. Furthermore, the replacement of Mo back contact with higher reflectivity and conductivity materials, such as Mo-Cu alloy, can be used to decrease the optical and electrical losses within the solar cell. Finally, the development of tandem CIGS solar cells has attracted the last decade the attention of the solar cell designers in the direction of overcoming the Shockley–Queisser limit of single-junction devices. As previously discussed, CIGS is an excellent material for developing tandem cells because of its tunable bandgap by altering the composition ratios [34]. However, the records up to now have shown that it is difficult to attain high-efficiency CIGS/CIGS tandems because of manufacturing reasons, such as the deterioration of the sub-cell characteristics during the development of the top cell. Nevertheless, numerous other types of materials can be combined with CIGS, such as other thin-film-based solar cell materials (e.g., $CuGaSe_2$ and CdZnTe), providing new prospects for improvements in the field.

Among the commercialized thin-film solar cells, CdTe technology presents the highest market share, accounting for around 7% at a global level [36]. CdTe has been demonstrated as the most reproducible and scalable solar cell technology so far, as evidenced by successful industrial manufacturing over the previous decade [30]. The key benefit of the under-consideration technology is the CdTe bandgap, which is considered ideal for the development of highly efficient single-junction solar cells. CdTe has a bandgap of roughly 1.5 eV in its single-crystal form and 1.45 eV in its poly-crystalline form. In principle, CdTe can deliver about 1 V under open-circuit conditions and 30 mA/cm^2 under short-circuit conditions, providing a promising way for developing solar cells with a maximum potential PCE of more than 27% [37]. Today, the efficiency records of these devices are at the level of 22% and 19.5% for small solar cells and modules, respectively, which are considered among the uppermost for thin-film solar cell technologies [3]. CdTe technology utilizes high-throughput manufacturing, although it requires high processing temperatures in the 600 °C range. They do, however, have the lowest module prices of any commercial solar cell technology today. Environmental concerns, such as cadmium toxicity and tellurium scarcity, have motivated research into alternate material systems that use non-toxic and abundant elements, with comparable ease of manufacturing [26].

The most common CdTe solar cells are developed on the basis of a p-n heterojunction structure, containing a p-doped CdTe absorbent layer and an n-doped cadmium sulfide (CdS) or magnesium zinc oxide (MZO) window layer. The CdTe absorbent layer is usually fabricated on top of a TCO-coated substrate. The devices are completed using a back metal contact, typically an Au or Ag contact, or a double-layered system composed of zinc telluride (ZnTe) followed by the metal layer or a carbon layer usually doped with copper (Cu) [38]. Both rigid and flexible CdTe solar cells perform better when developed in a superstrate configuration. This is happening for two main reasons [30]. The first one regards the so-called "activation procedure", which involves depositing a thin coating of cadmium chloride ($CdCl_2$) on the top of the CdTe and then annealing the stack in the air. This procedure, which could not be carried out in the case of substrate configuration, promotes recrystallization of the CdTe and enhances grains growth. It also improves the interface between the CdS and CdTe by enhancing the interdiffusion of sulfur into the CdTe and reducing the lattice mismatch between the two layers. The second reason is related to the back contact fabrication. For optimum performance, the back contact generally requires the addition of Cu. If Cu is deposited on the back of the solar cell fabricated in a substrate configuration, the Cu will interact with the CdTe layer during high-temperature processing, causing deterioration of the characteristics of the film. Nevertheless, recent studies are focused on overcoming this limitation by depositing Cu-free back contacts, such as MoO_x [30].

Current projects seek higher efficiency CdTe solar cells by increasing crystal quality and minority carrier lifetime, as well as improving doping control. Since the current response is considered at a satisfactory level in highly efficient CdTe solar cells, where the optical losses at short wavelengths are little and the cutoff at the bandgap is quite sharp, the main efforts today are put into increasing the open-circuit photo-voltage. In 2016, a research group attained the development of CdTe solar cell devices that exhibit open-circuit photo-voltage values higher than 1 V. This was achieved by doping CdTe with group V elements [39]. A noteworthy alternative strategy was accomplished by another research group the same year, which achieved breaking the 1 V barrier using a mono-crystalline CdTe/MgCdTe double-heterostructure solar cell. They demonstrated that the lattice-matched MgCdTe barrier layers provided excellent passivation to the CdTe absorbent layer, greatly increasing the minority carrier lifetime [40].

Fig. 4.12 Flexible thin-film solar cells application in real life [41–44]. (Reproduced from Ref. [41], https://en.wikipedia.org/wiki/Thin-film_solar_cell#/media/File:Thin_Film_Flexible_Solar_PV_Installation_2.JPG, under the terms of the CC BY-SA 3.0 license)

Comparison of Flexible Hydrogenated Amorphous Silicon, Copper Indium Gallium Diselenide and Cadmium Telluride Solar Cells

The flexible thin-film solar panels of today demonstrate a performance that rivals that of conventional rigid silicon panels, while their low-cost, bendable structure, lightweight and non-penetrating peel-and-stick installation set them ideal for a wide range of new and niche applications. Flexible modules can be applied from roofs of buildings and automobiles to sunshades and charge power banks for electronics (e.g., laptops, mobile phones, sensors), even in clothing accessories (see Fig. 4.12).

Flexible a-Si:H, CIGS and CdTe solar cells have received worldwide attention in our days since these technologies are already at a commercial level. For achieving high efficiency, CIGS and CdTe are more suitable candidates than a-Si:H technology. From them, CIGS usually attain higher performance than CdTe when developed to flexible foils due to the fact that the former technology is mainly developed in a substrate configuration, which is considered a better structure for developing flexible PV devices. In terms of stability, CIGS and CdTe solar cells are considered more robust to degradation compared to a-Si:H ones. Upscaling has been demonstrated for all thin-film solar cell technologies, although some limitations appear on the type of substrates employed. In terms of stoichiometry, producing a-Si:H and CdTe solar cells is easier than CIGS solar cells, with the CdTe solar cells having the simplest stoichiometry that helps to improve industrial scalability. On the other hand, a-Si:H technology is indicated for a wide range of applications because of its low-cost and long single-module manufacturing. For this technology, one of the most important advantages is its ability to be developed in a large number of different types of substrates. Thus, depending on the final application, production energy demand and installation space, the solar cell technology with the most suitable characteristics can be employed [30].

4.4 Single-Junction Gallium Arsenide and III–V Multi-junction Solar Cells

Although the energy produced by c-Si solar cells becomes more and more affordable, the PCE of these devices has reached a plateau, prompting scientists and engineers to ponder over alternative solar cell materials and architectures that can result in a more efficient light-to-electricity conversion. Gallium arsenide (GaAs) is one of the material choices for developing high-efficiency solar cells. GaAs is a nearly ideal material for solar cells application, with well-matched to solar spectrum bandgap, high optical absorption coefficient and extremely low non-radiative energy loss [4]. These characteristics set GaAs as an excellent candidate material for developing solar cells with substantially better performance than c-Si solar cells. GaAs solar cells now hold the global record for single-junction devices, achieving 29.1% by small-sized lab cells and 25.1% by small-sized modules [3].

On the other hand, multi-junction (MJ) designs are gaining significant interest in our days in the direction of achieving an even higher light-to-electricity conversion (see Fig. 4.13). MJ devices were early investigated in the decade of 1970s, with these solar cells employing elements from group III (aluminum, gallium, indium) and group IV (antimony, arsenic, nitrogen, phosphorus) of the periodic table [45]. The first MJ devices were dual-junction solar cells made up of an aluminum gallium arsenide junction stacked or grown on the top of a GaAs junction and interconnected by a semiconductor tunnel junction. To absorb sunlight effectively and avoid thermalization losses, III–V MJ solar cells often utilize a stack of two or more single-junction cells made of different bandgap materials [4]. The bandgap values of the solar cell materials are determined by the number of junctions utilized. Research has shown that it is possible to appreciably increase the PCE of these devices by increasing the junctions number, with a range of dissimilar bandgap materials matched to the range of photon energies presented in the solar spectrum [26]. According to theoretical models, the maximum efficiency of these devices can reach 72% [47]. Today, MJ solar cells fabricated by III–V semiconductors are the best-performing devices of all types of solar cells, demonstrating a PCE of 39.5%, while they also provide a series of important extra features, such as high radiation and thermal robustness, and high power-to-weight ratios [3, 48]. By the employment of concentrator solar cell designs, where lenses or curved mirrors are used to focus the sunlight onto a small area, this level of performance is further increased. Depending on the concentrator type, i.e., the amount of the solar concentration (measured in suns), these systems are categorized into low-concentration PVs (concentration of 2–100 suns), medium-concentration PVs (100–300 suns) and high-concentration PVs (concentration of 1000 suns or more) [49]. Today, the record PCE of concentrator MJ solar cells is

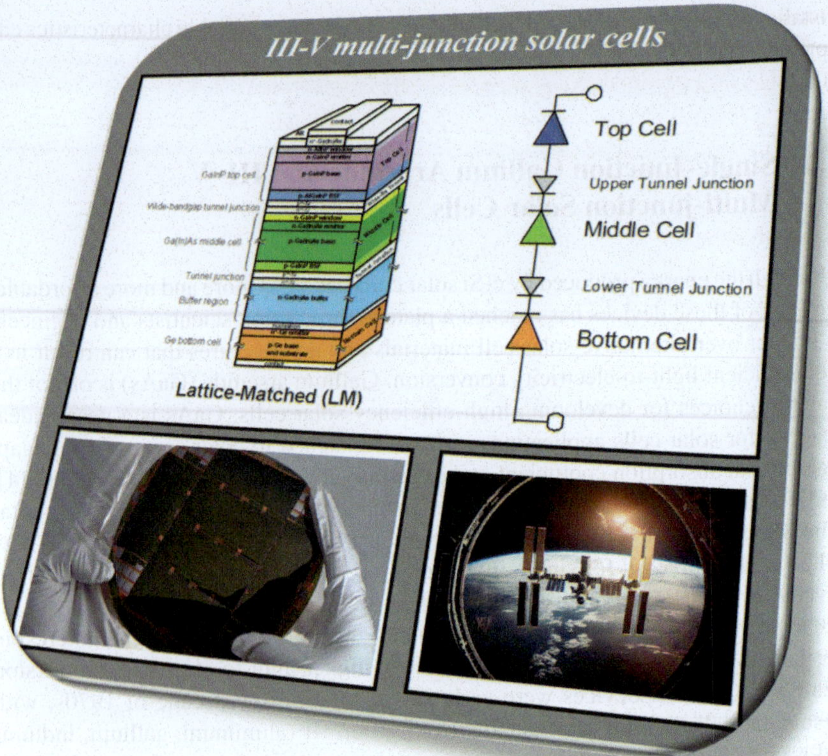

Fig. 4.13 Physical schematic of monolithic triple-junction solar cell and the corresponding equivalent electrical circuit diagram (top), image of a multi-junction solar cell and their application in the international space station (down) [45, 46]. (The image showing the international space station is reproduced from freepik.com, <a href = “https://www.freepik.com/free-photo/space-travel-vehicle-orbiting-planet-with-technology-generativeai_40933010.htm#query=space%20s tation&position=16&from_view=search&track=ais”> Image by vecstock </a> on Freepik, free license after attribution)

demonstrated as the record value of all types of solar cells developed throughout the history of the PV field, determined at 47.1% [3].

As it is perceived, III–IV compound materials developed in MJ designs provide a unique and reliable pathway for developing highly efficient solar cell devices. Nevertheless, the development of the under consideration solar cell technologies is yet highly restricted. This arises mainly from their high manufacturing cost. The cost of developing these materials is extremely high because of the need to achieve ultra-high-quality structures combined with their epitaxial growth, where the development of crystal defects and impurities highly reduces the performance of these devices. Thus, low-cost deposition techniques that lead to low-quality materials are generally unacceptable, subsequently leading to exorbitant costs, with the values often exceeding even $150/W. Considering that this is roughly 400 times the current prices of the energy produced by the mainstream c-Si solar cells, the applications of the under-consideration solar cell technologies are highly restricted to niche applications (such as space communications), where high performance, radiation and thermal robustness, as well as power-to-weight rations, are of great importance [50].

Today, key challenges for further developing III-V MJ solar cell technologies include optimization of solar cell architecture for different operating conditions, further improvement of their long-term stability, development of more efficient concentrator systems, and, of course, reduction of manufacturing cost without sacrificing device performance. Focusing on the latter topic, which is an important attribute for the limited application range of this group of solar cell technologies, several pathways have been proposed nowadays to deal with it [50]. The most obvious but also challenging way simultaneously is to improve production yield and volume. Increasing production volume can drive cost reductions via economies-of-scale and learning-by-doing tactics. This is an approach that has already been applied in c-Si solar cell technology and has led to a massive reduction in manufacturing costs. According to economic models developed for the III–V MJ solar cell technologies, cost reductions can be achieved by reductions in chemo-mechanical polishing costs, trimethylgallium and trimethylindium precursor prices, and substrate costs as production volumes increase, while MOCVD tool costs can also be reduced. Other cost reductions can arise from improved tool utilization and increased automation, and thus lower labor. According to estimates, the manufacturing costs of GaAs PVs can reach the level of $10/W if the production volume is set at 50 MW/year. Nevertheless, even though the cost of this order has been tolerable in certain markets, this is not compatible with the broad deployment of PVs in the widespread terrestrial market and additional strategies should be adopted. In this direction, the decrease of epitaxial growth costs is important. This can be achieved by increasing materials growth rate and utilization by increasing batch sizes and use of continuous, in-line machines, while additional reductions would require lowering material costs (such as by increasing the material utilization for the trimethyl precursors and the arsine and phosphine). Lowering the substrate costs is also essential to the aim of III–V solar cell cost reduction. This can be achieved by increasing the number of substrate reuses and subsequently reducing the polishing cost per reuse by using cheaper substrates than GaAs or Ge (replacement with lattice-matched or even with lattice-mismatched materials) and so on. Applying lower-cost metallization can also lead to further III–V solar cell cost reduction, such as using lower-cost fabrication techniques, e.g., electroplating or screen printing. These techniques are successfully employed in commercialized c-Si solar cell technologies. Putting all these measures together, the forecast production cost of III–V solar cells would likely be reduced to values on the order of $1/W or even lower.

4.5 Emerging Solar Cell Technologies

To move beyond the conventional and niche PV applications and successfully compete with the incumbent solar cell technologies requires emerging approaches that can lead to a higher performance-to-cost ratio and improved stability, as well as to the development of new PV devices that can provide new and unique features for light-to-electricity conversion. Shedding light on the future of PV systems, dye-sensitized solar cells (DSSCs), organic photovoltaics (OPVs), kesterite-based solar cells, quantum dot solar cells (QDSCs) and perovskite solar cells (PSCs) are the main emerging solar cell technologies developed the last past years (see Fig. 4.14). These PV technologies have emerged as a result of the high progress in materials science and devices engineering and are now in the early stages of research & development, and commercialization. In the majority of these devices, nanotechnology is employed to attain the requisite electrical and optical material characteristics for a variety of pioneer and alternative niche applications, including building integration, indoor applications, agriculture, automotive, wearable electronics, and so on.

Dye-Sensitized Solar Cells

Of the emerging solar cell technologies, DSSCs are among the most mature solar cells employing nanotechnology. The history of DSSCs begins back in 1988, when Brian O'Regan and Michael Grätzel co-invented these devices. The aforementioned scientists, three years later, demonstrated the development of the first high-efficiency DSSC, which could attain 7.9 and 12% light-to-electricity conversion under simulated sunlight and diffuse daylight conditions, respectively [51]. Because of their simple structure and easy fabrication, combined with their low manufacturing cost, satisfactory efficiency and stability, these devices early made practical applications seem attainable. Today, DSSCs demonstrate efficiencies of up to 14% (certified 13%

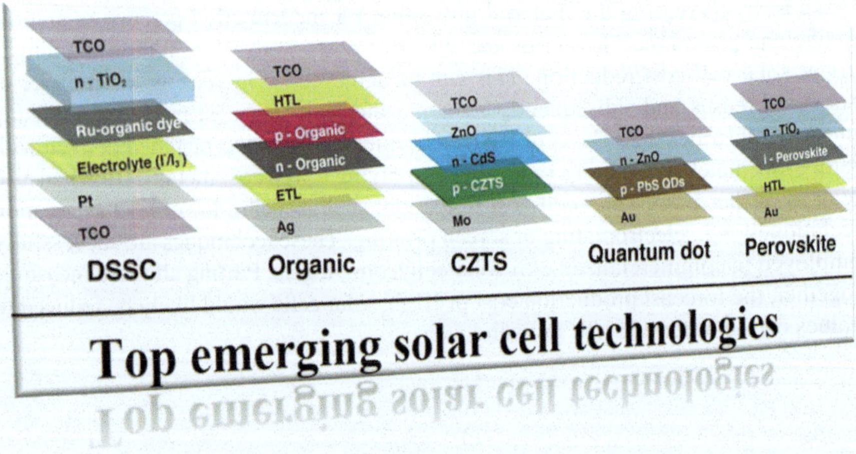

Fig. 4.14 Top emerging solar cell technologies [4]

Fig. 4.15 Some of the potential applications of dye-sensitized solar cells [57–60]

for small-sized solar cells and 10.7 and 8.8% for mini-modules and sub-modules, respectively) [52, 53]. Worldwide, but especially in Europe, Asia and Australia, dozens of research institutes and enterprises have been involved with the development, manufacturing and commercialization of DSSCs technology. As may be seen in Fig. 4.15, several different products have already been developed for entering the PV market. G24 Innovations introduced the first DSSC device to the market in 2009. Due to their flexible structure, these devices are able to be incorporated into a wide variety of applications, from bags and backpacks to wireless keyboards and other indoor electronic systems, allowing for convenient and on-the-go powering of consumer devices. 3GSolar is another company that develops dye-sensitized solar modules for usage in remote areas to provide power for lighting and irrigation pumps. Except for the aforementioned, up until today, many other companies have branches that are devoted to DSSCs development as well, providing prototypes for both outdoor and indoor applications. Some of these are Sony, Sharp and Aisin Seiki (in collaboration with Toyota Central R&D Laboratories) [54]. Key benefits arising from the usage of DSSC technology are their versatile and low-cost manufacturing, the usage of readily available and non-toxic materials, their high performance in diffuse light conditions, the flexibility of their design and their appealing aesthetics [55]. Nevertheless, the commercialization of DSSC technology is still hindered by a series of major challenges that the scientific community and solar cell developers are still working to overcome. Among these are their moderate light-to-electricity conversion under full sunlight, the long-term instability issues that downgrade their reliability for various applications, as well as the rigid structure and the liquid state materials that are employed in their conventional design, which limit their widespread application [56].

The most important attribute that is investigated today to be improved in DSSCs for their successful entry into the huge electricity production market is their PCE. Single-junction PVs have a maximum theoretical efficiency (Shockley-Queisser limit) of roughly 32% [61]. The largest divergence from this limit in inorganic solar cells is attributed to the loss-in-potential, which is defined as the difference between the optical bandgap of the light-active material and the open-circuit voltage. The loss-in-potential for high-efficiency c-Si solar cells is at the level of 400 mV, while the corresponding value for high-efficiency GaAs devices is at the level of 300 mV [62]. Unlike inorganic solar cells, DSSCs need relatively large over-potentials to operate. More specifically, for the conventional DSSCs that employ a TiO_2 semiconductor, a Ru metal complex dye and an iodide-based electrolyte, a 150 mV potential difference between the lowest unoccupied molecular orbital (LUMO) level of dye and the conduction band of TiO_2 is needed for a fast electrons injection and about 600 mV potential difference between the highest occupied molecular orbital (HOMO) level of the dye and the I^-/I_3^- redox potential for efficient regeneration of the dye. Namely, the loss-in-potential in typical DSSC systems is calculated at about 750 mV, which is a number greatly limiting the performance of these devices (PCE up to 13.8%) [56, 62]. In the last two decades, there has been a growing number of research efforts to exceed this limit by developing and applying novel materials and device designs. To this aim, the dye-sensitized working electrode, which constitutes the photo-active material system in the conventional n-type photo-electrochemical cells, is the most important part of the device for modification. At this part of the solar cell, there are demonstrated several options for boosting the light-to-electricity conversion. Some of them are the development of novel nano-architectures, semiconductor doping, fabrication of composites, interfacial engineering, increasing the light scattering and co-sensitization method [63, 64]. On the other hand, the development and optimization of new electrolytes also lead to the improvement of the PCE of DSSCs. The characteristics of the electrolyte have a significant impact on the light-to-electricity conversion by DSSCs, such as its conductivity and redox potential, its optical transparency, as well as electrolyte interaction with the electrodes [65]. For this case, the development of electrolytes with new effective redox mediators and the subsequent optimization of their synthesis (solvent, redox couple concentration, additives, etc.) is a promising strategy that can decrease the loss-in-potential and improve the optical characteristics of the system. Furthermore, the counter electrode has an important task in DSSCs operation, which is to catalyze electrolyte reduction. As in the former cases, the modification/replacement of this part of the cell, namely modification/replacement of the conventional Pt structures by novel nanostructures, can further improve the performance of DSSCs. For this case, the development and application of composite and hybrid nanomaterials, as well as of novel 3-D nanostructures (Pt-free or not), can improve the catalytic activity for electrolyte reduction and subsequently lead to an increased PCE by these photo-electrochemical cells [66, 67].

On the other hand, improving the stability of DSSCs, at molecular, cellular and module levels is imperative for the intended use of these solar cells in a wide range of practical applications. Each of these scales poses unique scientific and technical challenges that should be overcome to decrease the instability issues presented in these devices [54]. In the last decade, there has been a rising amount of research efforts to achieve the stability criteria for various PV applications. These are usually evaluated using accelerated ageing testing under various environmental conditions, according to the standards and protocols established between researchers in the field [68]. Herein, the most important component in DSSCs that should be considered for modification/replacement is the liquid state electrolyte. DSSCs have made significant progress in terms of efficiency through the use of liquid state electrolytes. Nevertheless, the stability of these devices is deemed inadequate for long-term operation [65]. Major factors leading to the progressive decrease of the performance of DSSCs are alterations in the characteristics of electrolytes and the anode/electrolyte interface [56]. In this direction, selecting the appropriate solvent, redox couple concentration, electrolyte cations and additives is crucial for increasing solar cells long-term durability. Novel liquid state electrolytes that are engineered in terms of their composition to demonstrate stable DSSCs devices are continuously investigated. On the other hand, the replacement of the liquid state electrolytes with quasi-solid state electrolytes or solid state hole transport materials is considered today the most effective strategy for increasing DSSCs reliability for various applications [65]. Although the performance of DSSCs using the aforementioned materials is usually lower than that of conventional devices, they are viable alternatives with enhanced stability and sealing ability. Today, the scientific community has focused on the development of advanced alternatives to the commonly used liquid state electrolytes of DSSCs. These include polymer electrolytes, nanocomposite electrolytes and polymer composite electrolytes [65]. These materials possess both the diffusive property of a liquid and the cohesive property of a solid, providing better long-term stability than liquid state electrolytes and merits of liquid state electrolytes, including high ionic conductivity and excellent interfacial contact properties. In this field of research, novel charge transfer mechanisms and strategies to attain high-quality interfaces in the multilayered solar cell structure have been intensively investigated in the last years [69–71]. The development of solid state hole transport materials is also investigated, both in the cases of inorganic and organic p-type materials (doped and undoped) [65]. On the other hand, to stabilize the performance of DSSCs, several scientific efforts are put into the development of a more stable dye-sensitized working electrode, in terms of no energy levels shifting, surface states increment and degradation/desorption of the dye-sensitizer [72]. In this direction, the main strategies include the anode surface modification/post-treatment, semiconductor doping, development of new dyes, etc. [73–75]. Finally, to improve the stability of DSSCs, modifications are made to the counter electrodes. Degradation of the counter electrode of DSSCs can lead to an increased charge transfer resistance at the counter electrode/electrolyte interface and increase the recombination rate inside the solar cells [76]. Up until today, there have been many scientific efforts focusing on the stability of counter electrodes of DSSCs. These include the development and application of metal-based alloys, carbon-based nanomaterials, polymeric nanocomposites and so on [77–79].

Increasing the application range of DSSCs by providing unique device characteristics without sacrificing their performance and stability, is also a topic of research and development in the last decades by DSSCs developers. DSSC devices are continuously improved to operate better in diffuse light and low light conditions, or even in artificial light, in order to be able to be used in new PV applications, such as consumer electronics. To this aim, modifications in dye-sensitized working electrodes and electrolytes are the most common. These include the development of new dye-sensitizers, novel anode semiconductor nanostructures and electrolytes that can provide the appropriate optical and electrical characteristics for this group of applications [80, 81]. On the other hand, flexible DSSCs have already been introduced into the global lifestyle product market, and efforts are now put into the development of more efficient and stable flexible devices. In this direction, the replacement of the conventional rigid glass with alternative substrates that can provide flexibility, lightweight and/or mechanical robustness is of first priority for investigation [82, 83]. Of course, for this kind of modification, alterations in the manufacturing processing of the electrodes of DSSCs are needed to maintain a high PCE. Additionally, the replacement of liquid state electrolytes with quasi-solid state electrolytes or hole transport materials facilitates the increased reliability of the devices from leakage/evaporation failures [65]. The achievement of spectral-engineered and transparent DSSC devices has also been a topic of concern for the scientific community over the last decade. Under the appropriate modifications of the DSSCs structure, such as increasing the transparency of working and counter electrode, as well as of the electrolyte, combined with the design of new dye-sensitizers that can absorb sunlight effectively in specific spectral ranges while leaving the rest part of sunlight to pass through, new and niche applications for DSSCs appear [84]. The ability to achieve appealing aesthetics in these devices, retaining at the same time a satisfactory performance, is also a matter of investigation. Research institutes and enterprises are working on the development of multi-colored DSSCs by the usage of the appropriate manufacturing techniques that can allow the development of complex, multi-layered patterns and color density gradients on the electrodes [85]. Mechanical robustness in solar cells is also a feature that has seized the attention of the scientific community today in the direction of developing structural solar cells that are capable of producing electric power from sunlight while possessing mechanical integrity. Metal-based, textile-based and fiber-based DSSCs are indicated for the development of mechanical robust devices, which can undergo complex mechanical loadings and deformations, without deterioration of their performance [86–88].

Finally, reducing further the cost of DSSC devices is also a matter of investigation today, given the continuous decrease in the cost of electricity produced by commercially available solar cells. Taking into consideration that the most expensive parts in a DSSC structure are the TCO substrate, the Pt-based cathode and the dye-sensitizer, several modifications of the conventional electrodes of these solar cells are studied with the aim of decreasing the manufacturing cost of DSSCs without deteriorating their performance. In the past decades, many scientists have been devoted to the fabrication of lower-cost dye-sensitized working electrodes and counter electrodes for DSSCs than their conventional counterparts. The replacement of the conventional TCO glass with lower-cost surface conductive substrates is of first priority [82]. The usage of surface conductive plastics is a primary way to reduce the manufacturing cost of DSSCs, in terms of the usage of cheaper substrates and the possibility of applying lower-cost and high-throughput fabrication techniques, such as roll-to-roll production. Furthermore, the replacement of the conventional TCO coating with lower-cost alternatives, such as ultra-thin carbon-based nanostructured films and metal wire meshes, is a strategy that has often appeared in recent years [89, 90]. The usage of low-cost metal foils is also investigated by scientists in different DSSC designs [91]. On the other hand, during the last decades, several research efforts have been made for the development of cost-effective counter electrodes for DSSCs that could maintain the light-to-electricity conversion efficiency at high levels. These mainly relate to the development of novel high performance Pt nanostructured electrodes that are characterized by a reduced amount of the costly Pt, but at the same time, they provide a satisfactory catalytic activity. On the other hand, given that Pt is a costly noble metal, reasonable efforts have been made to produce cheaper alternative cathode materials [66, 67]. These include carbonaceous materials, conductive polymers and transition metals. Regarding the dye-sensitizer, research on the synthesis of novel organic dye complexes and the development of new dye sensitization methods are also applied, aiming at lower-cost DSSCs [85, 92]. Finally, the cost of the solvent-based electrolytes is not considered as high for DSSCs wide commercialization. Nevertheless, the development of advanced quasi-solid state electrolytes and solid state hole transport materials is still costly in many cases and should be replaced/modified by lower-cost alternatives. In this case, the investigations are focused on the development of lower-cost ionic liquid electrolytes and polymer electrolytes from cheap polymeric molecules, as well as of hole transport materials that are synthetically cost-effective [93–96]. Finally, the adoption of up-scalable and high throughput fabrication techniques can provide a way for more competitive DSSC products [97].

Organic Photovoltaics

Following almost twenty years of research and development, OPV technology nowadays experiences great progress toward its consolidation into the PV market. The record PCE of these devices has greatly increased recently to 18% for lab-scale devices, while devices on the module level can now achieve efficiencies up to 12% [3]. Combined with the potential for low-cost solar-to-electricity conversion, mechanical flexibility, lightweight, appealing aesthetics, absorption tunability, environmental friendliness and ease of scale-up to an industrial level, these solar cells are considered one of the most important emerging technologies in the field [98]. OPVs are mainly fabricated by thin layers of semiconducting organic materials (polymers or oligomers) that can absorb photons from solar radiation effectively to convert them into electric power using low-cost manufacturing (e.g., roll-to-roll processing). Today, OPVs are candidates for a number of new and established applications, such

Fig. 4.16 Roll-to-roll printed organic photovoltaic (left), the versatile applications of organic solar cells (right) [98]. (Left is reproduced after permission given by "infinityPV ApS" and right is reproduced from Ref. [98], https://ars.els-cdn.com/content/image/1-s2.0-S2666935821001026-ga1_lrg.jpg, under the terms of the CC BY-NC-ND 4.0 license)

as building integration, indoor PV energy production and wearable electronics (see Fig. 4.16) [98].

One of the key advantages of OPVs over inorganic solar cells is their amazing mechanical characteristics that demonstrate, which enable their use in many unprecedented solar cell applications. Due to the intrinsic flexibility of organic semiconductor materials employed in these solar cells, efforts have been increasingly devoted to the development of high-performance flexible devices using flexible transparent electrodes [99]. Moving forward from the conventional TCO-coated plastics, more and more efforts are being plunged today into the development of advanced alternatives, such as conducting polymers, carbonaceous materials, metal meshes and hybrids. On the other hand, due to their potential usage in building-integrated PV applications, transparency is a feature that is gaining more and more attention [100]. Organic semiconductors provide superior tunability of their optical bandgap compared to inorganic materials, a characteristic that can contribute to their spectral engineering easily. Moreover, this group of materials demonstrates high absorption coefficients, which can enable sufficient light absorption even by the use of ultra-thin films. Furthermore, opportunities for the rapid development of indoor OPVs have arisen, given the proliferation of low-power electronic devices enabled by the rise of the Internet of Things and their corresponding reduction in power consumption and cost. Given that c-Si solar cells absorption characteristics do not match the spectrum range of the majority of daily used indoor light sources, while solar cells based on III–IV semiconductors generally require very expensive manufacturing, OPVs have great potential in the indoor PV market. Yet, major setbacks still limit their widespread commercialization, mainly related to their poor long-term durability under sunlight and non-competitive PCE compared to the commercialized solar cell technologies [98]. Table 4.3 presents

Table 4.3 Advantages and limitations of organic photovoltaics technology in the development of flexible, semitransparent and indoor solar cells

Organic solar cells	Advantages	Limitations
Flexible	Superb mechanical characteristics	Low performance
Semitransparent	Ease of spectral engineering	High compromise between visible transmittance and device efficiency
Indoor	Compatible optical characteristics with indoor light sources High absorption coefficients Low effect of series parasitic resistance	High energy loss due to charge carriers recombination

the main advantages and current challenges of OPVs technology for the development of flexible, semitransparent and indoor solar cells.

The inferior OPVs stability remains the most critical challenge before this technology enters the commercial arena. Metastable morphology, diffusion of developed impurities from the electrode and buffer layer into the main active layer with time, degradation due to irradiation, oxygen, water, heating and mechanical stress are some of the factors limiting the lifetime of these devices [101]. More specifically, metastability is a feature that characterizes the core materials of these solar cells, which is generally composed of two or three phases (the donor phase, the acceptor phase and the donor/acceptor mixed phase). Phase separation is usually metastable in these systems due to the strong mobility of organic materials. Higher instability in the morphology of the active layers also appears with the usage of solvent additives in these devices. Furthermore, like the donor and acceptor materials, the electrodes and buffer layers also show mobility. Impurities developed in time by the degradation of these layers usually diffuse into the main active materials and reduce the performance of the solar cells by changing the energy levels of the multi-layered structure, acting simultaneously as traps for charge recombination. On the other hand, OPVs also present instability issues upon continuous solar irradiation. Upon irradiation, photo-physical and photo-chemical deterioration in the active layer, buffer layer and active layer/electrode interfaces are the original causes of instability in these devices. In the active layer, the photo-oxidation process is also responsible for the majority of the photo-chemical deterioration. Oxygen can also reduce the stability of OPVs, even in the dark. At first, the metal electrodes with a low work function can be oxidized by oxygen permeation. Secondly, the penetrative oxygen in the active layer takes part in various photo-oxidation reactions of donor and acceptor materials. Thirdly, the oxygen doping in the active layer can enhance hole concentration, which in turn results in an increased density of deeper traps for electrons and a decreased PCE. The ingress of water in the device is another important factor limiting the stability of OPVs. Water can diffuse into the devices via hygroscopic materials and metal electrodes, causing the development of defects, which in turn act as pinholes that allow more water ingress. When the water molecules reach the active layers, they can induce materials aggregation and decrease the donor/acceptor interface area, subsequently reducing the performance of the devices. Additionally, these solar cells degrade during their operation under high-temperature conditions. Heating-induced physical degradation frequently appears in these devices, particularly for temperatures higher than the glass transition temperature (T_g) of the applied materials. The enhanced mobility of polymers above their T_g not only causes aggregation states in these systems but also assists alterations in their crystal structure that deteriorate their optoelectrical characteristics. Finally, the mechanical stress and plastic deformations also lead to the degradation of the PCE of OPVs. Delamination and fracture failures can appear during mechanical stressing of these devices, which reduce their proper operation, while also provide ways for additional degradation, such as by oxygen and water ingress.

To deal with the aforementioned instability issues and further increase the reliability of OPVs for extended applications, many strategies have been proposed and implemented up until today. Active layer material re-design (including donor molecular skeletons and side chains, as well as fullerene and non-fullerene acceptors), active layer device engineering (including ternary blends and processing methods), active layer inversion, buffer layer optimization (including hole and electron transport layers), development of higher stability electrodes (including transparent and back electrodes) and new hermetic encapsulation methods are some of the main approaches [101]. More specifically, to increase photo and photo/air stability of OPVs, many studies propose the employment of cleavable side chains and photo/air stable units on the polymer, the increment of polymer crystallinity and decrease of fullerene crystallinity, the decrease of defects/traps in the active layer, as well as decrease oxygen permeation to the active layer. On the other hand, to increase the air stability, the decrease oxygen permeation and water ingress into the active layer, the use of electrodes that exhibit low reactivity with oxygen and water, the usage of charge transport layers with good inoxidizability, low hydroscopicity and neutral PH, and modifications on the conventional device geometry are proposed. To increase the thermal stability of OPVs, the main strategies are focused on the decrease of fullerene crystallinity, the usage of polymers characterized by higher T_g, the employment of cross-linkable polymers/fullerenes or addition of cross-linkers, the removal of residual solvents in the active layer, and the formation of more stable donor/acceptor and active layer/electrode interfaces are investigated [101].

The second main challenge that needs to be tackled before the widespread commercialization of OPVs is their moderate performance. Although the performance of OPVs has rapidly increased in the last few years, demonstrating PCEs on the order of 18%, higher efficiencies is still needed to commercialize this emerging technology. The light harvesting and collection of the photo-generated charge carriers in these devices are limited by multiple factors, which are related to both materials properties and device engineering. Major obstacles to further increasing their PCE are the poor light absorption, especially at wavelength ranges out of the visible spectrum of light, the non-optimized energy levels of the multi-layered structure, as well as the recombination losses of the photo-generated charge carriers [102].

In the direction of increasing OPVs performance, the emergence of non-fullerene acceptors as alternatives to the commonly employed up to now fullerene-based materials has recently revitalized the field. These materials are developed using facile synthesis methods, while they demonstrate strong light absorption and readily tunable energy levels. Non-fullerene acceptors are typically classified into two broad groups, namely small-molecule acceptors and polymeric acceptors [102]. Amongst them, fused-ring electron acceptors represent the most important small-molecule acceptors since their implementation in OPVs has shown great potential for the achievement of high PCEs (efficiencies over 18% have been demonstrated) [103, 104]. Today, the scientific community investigates the further improvement of the performance of non-fullerene-acceptor-based OPVs. To this aim, key aspects are the development of viable alternative molecular designs to achieve higher electron mobility and lower energy loss, better energy levels matching between the donor and acceptor materials to maximize the photo-voltages produced by these solar cells, while the achievement of preferred blend morphologies is quite important to achieve higher charges separation and transfer. Furthermore, the development of acceptor/donor systems of complementary absorption spectra is a promising strategy that can be employed to increase electric power production by utilizing as many photons as possible [102].

On the other hand, numerous studies have claimed that incorporating inorganic nanoparticles into OPVs structure is a successful technique for addressing the majority of the PCE limitations [105]. More specifically, with the introduction of nano-dopants, such as metal or semiconductor nanoparticles, the morphology of the active layer and, hence, the performance of the devices can be improved, while the incorporation of nanoparticles of high electric permittivity into the active layer has shown that can improve the dissociation of excitons through reducing the Coulomb attraction between the electrons and the holes. Furthermore, the incorporation of ferromagnetic nanoparticles has been proposed as an effective strategy to reduce recombination losses and enhance device efficiency in these devices. This is because the interaction between a weak magnetic field and singlet excitons induces the inter-system crossing effect, resulting in the conversion of singlet into long-lived triplet excitons. This is important because the lifetime and diffusion length of triplet-state excitons are orders of magnitude higher than those of their singlet-state counterparts. Thus, increasing the population of triplet excitons reduces recombination losses. On the other hand, the incorporation of metal nanoparticles into the active layer and/ or buffer layer has been shown to increase light absorption due to localized surface plasmon resonance (LSPR) and light scattering. Besides, to broaden the absorption spectrum of OPVs, both up- and down-converters based on rare earth metal nanoparticles have also been investigated. In this strategy, downconverters possess dual functions: through the conversion of the above-band gap photons (mainly UV) into visible light and subsequently increasing the PCE of the device, as well as by reducing the amount of UV reaching the core of the device, leading to increased devices stability under light-soaking. Instead, upconverters are used to convert sub-bandgap photons into visible light, also leading to increased PCE in the OPVs.

Finally, an alternative and highly promising approach to boost the OPVs efficiency is to use ternary, quaternary and multiple-component bulk heterojunctions [106]. This strategy can not only enhance the absorption coverage by using multiple materials with different bandgaps but also preserve the processing simplicity of the binary devices. Moreover, the incorporation of an additional component in the active layer can also play a crucial role in optimizing molecular structure, nanoscale morphology, long-range order and crystallinity. It can also be used to tune frontier molecular orbital levels by forming homogeneous donor or acceptor phases, which can provide a more effective charge transfer and lower energy losses. The ternary approach is particularly viable in systems that use non-fullerene acceptors, as these are typically designed to absorb well in the visible region and are easily tunable, meaning a system of components with complementary band gaps can be generated.

Kesterite Cu_2ZnSnS_4, $Cu_2ZnSnSe_4$ and $Cu_2ZnSn(S,Se)_4$ Solar Cells

The challenge of finding alternative solar cell materials based on elements that are non-toxic and naturally available in abundance on the Earth's crust prompted the exploration and development of solar cells based on Cu, Zn and Sn chalcogenide absorbers. Kesterite Cu_2ZnSnS_4 (CZTS) and its selenium alloys, i.e., $Cu_2ZnSnSe_4$ (CZTSe) and Cu_2ZnSn (S, Se)$_4$ (CZTSSe), have been fascinating the last years due inter alia to their high absorption coefficient (~10^4 cm^{-1}) and direct band gap of 1–1.5 eV that enable the absorbance of most of the visible solar spectrum [107]. Such solar cells have a potential theoretical PCE limit of around 32%. When these materials are combined with cadmium-free buffer layers, they provide a way for the development of completely non-toxic PV devices [108]. Up until today, several methods have been reported for the fabrication of kesterite-based solar cells. Even though few studies are published annually on this technology, many methods have been proposed and applied to synthesize this group of chalcogenide absorbers. These include spin-coating, hydrothermal processing, hot injection, electrochemical deposition, sputtering, chemical vapor deposition and others. Each method excels in different material characteristics, such as purity, grain size growth, porosity, etc. [109]. Today, the record PCE of kesterite-based solar cells is on the order of 13% [3].

Table 4.4 lists the characteristics of the CZTS, CZTSe and CZTSSe materials and of the corresponding kesterite-based solar cells. For comparison purposes, the characteristics of the chalcogenide CIGS absorber and the commercialized CIGS-based solar cells are also presented. As it is perceived, most of the characteristics of kesterite materials are similar to the corresponding ones of CIGS. Herein, the main difference between the former group of materials compared to the latter one is that kesterite materials demonstrate a very low charge carriers lifetime. This is a result of cation disorders defects and secondary phases that are usually developed during the synthesis of the quaternary compounds because of the difficulty in controlling their morphology, phase structure and stoichiometry. This is not usually encountered in ternary (CIGS), binary and unary materials since their synthesis is of lower complexity [107]. These defects alter the characteristics of the main active material locally. Namely, secondary phases are developed that behave differently, such as a

Table 4.4 Characteristics of the Cu_2ZnSnS_4, $Cu_2ZnSnSe_4$ and $Cu_2ZnSn(S, Se)_4$, copper indium gallium diselenide materials and correspondingly of the efficiency of the solar cells [107]

Characteristic	CZTS	CZTSe	CZTSSe	CIGS
Crystal structure	Tetragonal	Tetragonal	Tetragonal	Tetragonal
Lattice constants (a, c) Å	a = 5.427, b = 10.840	a = 5.40, b = 10.846	a = 5.40 b = 10.846	a = 5.6–5.8, c = 11.0–11.5
Bandgap (eV)	1.4–1.5	1.0	1.4–1.5	1.0–1.7
Absorption coefficient (cm^{-1})	10^4	10^4–10^5	10^4–10^5	10^4–10^5
Carrier concentration (cm^{-1})	10^{15}–10^{20}	10^{15}–10^{20}	10^{15}–10^{20}	10^{17-18}
Carriers lifetime majority/minority (ns)	15/2.5	15/2.5	15/2.5	250
Density (g/cm^3)	4.56	4.56	4.56	5.7
Melting point (°C)	990	990	990	1070–990
Efficiency (%)	8.4	10.1	12.6	22.6

metal-like, p/n-type semiconductor or insulator, highly increasing the recombination rate inside the devices and decreasing correspondingly the light-to-electricity conversion provided by the solar cells.

The PCE loss faced due to the high loss of charge carriers recombination is, up until today, the main challenge that has to be overcome in order to make this technology more competitive for entering the PV market. Today, several cutting-edge strategies are employed to further enhance the performance of kesterite-based solar cells. Modifications are applied to the absorbent layer, buffer layer, window layer and back contact (see Fig. 4.17). On the other hand, only a very few investigations have been reported on the stability of kesterite-based solar cells, both for indoor and outdoor applications, yet demonstrating that obtained reliability is not the major factor hindering these solar cells commercialization [110].

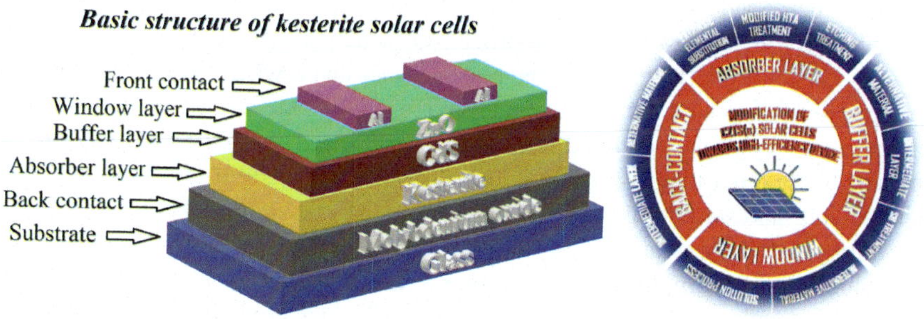

Fig. 4.17 The basic structure of kesterite-based solar cells and proposed modifications for increased light-to-electricity conversion [111]

Several efforts, including extrinsic element replacement, modified high-temperature annealing and etching treatment, can be used to improve the quality of the kesterite absorber layer [111]. Extrinsic elemental substitution is a promising method for bringing about changes in the electronic structure, bandgap energies, and even the crystallinity of kesterite materials. Up to now, several extrinsic elements, e.g., Ge, Sb, Na, Mg, Ag and Cd, have been studied as single or synergistic (two or more) external sources to improve the quality of the bulk kesterite absorber. Meanwhile, a kesterite absorber with graded bandgap energy across the absorber layer can be fabricated by incorporating different extrinsic elements on the top and bottom layers. A band shift with a lower bandgap towards the buffer layer and a higher bandgap towards the back contact layer can be obtained with this strategy. High-temperature annealing treatment can also be applied to improve the quality of the kesterite absorber. Herein, a multistep thermal treatment is usually applied, involving a lower holding temperature (typically at 300 °C) and a higher holding temperature (typically higher than 500 °C) to promote optimum kinetics of solvents removal and crystal growth. Moreover, the etching treatment is usually employed to remove the presence of surface contaminants, secondary phases and metal oxides.

On the other hand, the low PCE of kesterite-based solar cells arises from the non-optimal band alignment at the absorber/buffer interface. Although CdS is the most widely used buffer layer in these solar cells, undesirable cliff-like band alignment is often raised at the kesterite/CdS interface, which acts as an energy barrier, resulting in an increased recombination rate inside the solar cells [112]. Therefore, several efforts have been put today to reduce this characteristic by changing the alignment to a more desirable spike-like alignment. Herein, the use of interlayers at the absorber/buffer interface has been extensively investigated by using materials such as SnO_x, Al_2O_3, In_2S_3 and ZnS, acting also as passivation layers and shunt blockings. Alternatively, soft-baking treatment after buffer layer deposition has shown that it leads to a more favorable band alignment due to an interdiffusion phenomenon appearing at the absorber/buffer interface. Engineering the conventional window layer of kesterite-based solar cells can also increase the light-to-electricity conversion and the manufacturing cost of these devices. Commonly used TCO coatings, i.e., indium tin oxide and fluorine-doped tin oxide, are replaced by novel hybrids that can be developed by alternative lower-cost fabrication processes, which simultaneously lead to better charge transfer. Finally, regarding the back contact layer, where Mo is the standard material used up to now, the usage of interlayers made of metals, metal oxides, metal nitrides, metal sulfides, carbon, etc., has been reported to lead to PCE improvements. On the other hand, Mo alternatives are simultaneously studied for increased performance and stability in kesterite-based solar cell technology [111].

Amongst all these strategies, postdeposition thermal treatment for bare absorber, heterojunction and completed device, as well as alkali doping and bandgap grading by engineering graded cation and/or anion alloys, are considered cutting-edge options. These methods have already been applied to the CIGS solar cells, demonstrating step-change improvements in the PCE of these PVs, and are anticipated to push kesterite-based solar cells to new records of performance [113].

Quantum Dot Solar Cells

Reducing the size of structural units of semiconductors to dimensions equal to or less than twice the exciton Bohr radius enables tweaking the characteristics of the materials, with control of the bandgap being the most manifest feature (see Fig. 4.18) [114]. This is the so-called quantum confinement regime, and the materials exhibited it in the three dimensions are known as quantum dots (QDs). In the last decade, semiconductor QDs, especially colloidal and self-assembled, have drawn considerable attention in the PV field due to their unique optoelectronic advantages [115]. Both groups of materials have the potential to highly improve charge carriers photogeneration and subsequently enhance the PCE of solar cells than today's records. They are usually employed as the main active PV material in solar cells, given the advantage of processing their optical and electrical characteristics that can be tuned by simply changing the size of their structural units. This enables the easy spectral engineering of the solar cells, allowing for effective photon harvesting [4]. Today, the QDSCs demonstrate a record efficiency of 18.1% [3]. Nevertheless, the commercialization of this solar cell technology is still hindered by a number of major challenges, mainly related to the incomplete understanding of QDs chemistry, still leading to unsatisfactory performance and stability [116].

The earliest reports of solar cells employing QDs regard devices that use QDs as complementary materials of the main non-quantum-confined light active materials. The first report on PV devices using QDs as the primary absorber appeared in 1998, where InP-based QDs were applied as sensitizers in a DSSC configuration. Since then, several alternative architectures have appeared, aiming for better exploitation of the advantages arising from the employment of QDs in solar cells [119, 120]. Schottky QDSCs are devices first demonstrated in 2005, employing thin films of QDs for both absorption and charge carriers transport. These cells take advantage of the work function difference between the TCO and the top metal contact to generate a built-in field, which can subsequently lead to photo-current production. However, it was early understood that this architecture possesses serious limitations regarding the device performance, especially in terms of low charge carriers transportation and low photo-voltage production. To overcome the limitations associated with Schottky

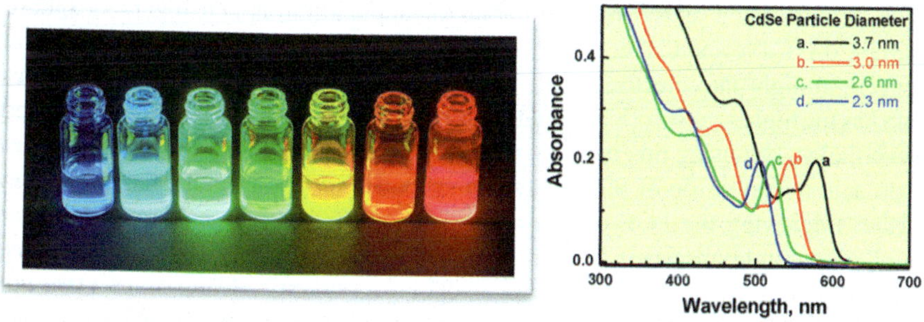

Fig. 4.18 "Perovskite quantum dots can express a wide range of colors", Canon Inc. (left) and absorption spectrum of different-sized CdSe particles (right) [117, 118]

QDSCs, heterojunction QDSCs were developed. Taking inspiration from sensitized solar cell architecture, in the heterojunction architecture, a highly doped n-type metal oxide is employed in combination with a p-type QD film to form a heterojunction. Light-to-electricity conversion begins, therefore, in regions close to the junction, overcoming one of the limitations of the Schottky architecture. The relative free carrier densities in the materials on each side of the junction determine the widths of the depletion regions. Typically, the wide-band-gap semiconductor is highly doped, with the result that the depletion region extends mostly over the QDs film. Drift currents created by the electric field are used to drive carriers in the QDs film. To this architecture, many modification strategies have been employed up until today to further increase the performance of solar cells. However, performance in heterojunction QDSCs based on planar films is limited by short carrier transport lengths, something that limits the thickness of the main active QDs layer and, subsequently, the light absorption capability of the devices. To overcome the absorption—extraction compromise, bulk heterojunction QDSCs are developed. In this architecture, the n-type wideband-gap semiconductor and QDs film form an interpenetrating layer. The structured interface makes it possible to increase absorption and carrier collection by extending the depletion zone and addition of more QD-structured material. Nevertheless, the increased interfacial area inherent to the bulk heterojunction design introduces the issue of increased bimolecular recombination. As a result of the necessity to control the offsets between the conduction bands of the QDs and the metal oxide, the device built-in voltage drops, which is still a major challenge in further enhancing the performance of QDSCs. Followed research in the field showed that the drawbacks associated with heterojunctions could be overcome by using homojunction-like architectures where both the p- and n-materials of the junction are composed of QDs. In this direction, quantum junction solar cells are demonstrated. Simultaneously, nano-heterojunction QDSCs appeared after the development of n-type nanocrystals. In these architectures, both materials can contribute to photo-charge generation. As charges were produced in both materials, the performance of solar cells is greatly enhanced compared to a bilayer structure. Subsequently, many additional architectures appeared to further improve the performance of QDSCs, such as quantum funnels, graded doping architectures and multiple-junction devices [120].

Among all types of QDs, metal chalcogenide QDs, especially Cd-based, have been demonstrated as the most suitable for application as sensitizers in QDSCs [121]. CdS, CdSe and CdTe, have been commonly employed as photo-sensitizers to convert sunlight into electricity due to their highly fluorescent nature. From these, CdS has the highest absorption limitations due to its relatively large bandgap of 2.4 eV. Nevertheless, by using this material combined with CdSe QDs in a co-sensitization approach, an extension of the absorption spectrum of a solar cell can be achieved, further improving the PCE of these devices compared to the corresponding ones employing single absorbers. Up until today, many additional strategies have been adopted to improve the performance of Cd-based QDSCs. Capping of QDs with suitable organic/inorganic ligands, post-synthesis treatments with ZnS, SiO_2, etc., doping/co-doping of QDs with transition metal ions (e.g., Mn^{2+}, Co^{2+}) and optimization of the synthetic procedure are some notable approaches that have been widely explored as means of extending the absorption spectrum and reducing charge carrier recombination inside the solar cells [121]. Simultaneously, carbon-based QDs have emerged as new promising alternatives to metal chalcogenide QDs for application in QDSCs due to their ease of synthesis and tunable photoluminescent characteristics. Yet, new strategies and protocols are still needed to improve their efficiency. Amongst others, metal halide perovskite QDs have recently drawn exceptional attention from the scientific community for their implementation as active materials in solar cells and are now considered superstar materials for light-to-electricity conversion [122]. Colloidal nanocrystals of lead halide perovskites offer additional benefits in the field of PVs because of their very high absorption coefficients, bandgap tunability and phase stability [120]. Furthermore, perovskites are often considered "defect tolerant", meaning that their inherent defects do not act as electron/hole trap states. When compared to the more common semiconductor QDs, such as those based on the II–VI, III–V or IV–VI groups, perovskite QDs stand out due to their naturally low trap density [123]. This low defect density renders electronic surface passivation with ligands or wider bandgap materials a less strict requirement. On the other hand, the exploitation of new types of electrolytes and hole transport materials has also been tested intensively over the last decade to overcome the performance bottleneck appearing in QDSCs. To this aim, cobalt-based electrolytes and p-type semiconductors, such as CuI and CuSCN, have appeared as promising candidates for application in QDSCs [124].

Up until today, over three hundred research papers have already been published, focusing on enhancing the stability of QDSCs. However, the physical and chemical reactions involved in these solar cells remain largely unknown, and their poor reliability is still an important challenge that needs to be tackled before their commercialization. Today, given that the topic of stability of QDSCs is of first priority, an ever-increasing number of effective strategies appear to deal with the appeared instability issues, focusing on both QDs, electrolyte/hole-transport-material and counter electrode components of these devices [125]. As the literature has shown, the development of core/shell QDs is one of the most attractive strategies to increase these devices durability by protecting the QDs from surrounding detrimental factors such as oxygen, moisture, heat, etc. On the other hand, the development of novel QDs, other than chalcogenides, is also investigated. The development of new innovative redox couple electrolytes with low oxidation potential, low surface tension and good hole transferring ability is also tested to minimize instability issues, such as QD desorption and chemical corrosion. Besides, the usage of quasi-solid state electrolytes in replacement of the conventional liquid state electrolytes can provide higher operational durability in these devices, minimizing events such as electrolyte leakage and evaporation. The development of suitable electrolyte—counter electrode systems is also explored as one of the most critical aspects for extending the lifetime of QD solar cells. Studies have shown that carbon-based nanomaterials with high conductivity, high-surface-area and porous structure can be a promising route to reduce corrosion-related instability issues. Moreover, the structural and chemical modification in carbon-based nanomaterials can be considered to improve further the PV characteristics. Finally, improving the encapsulation of the core of the solar cell devices is a rather effective strategy to maintain the performance of these devices at high levels throughout the time, stressed under different operational conditions. In order to achieve stable devices, the use of novel encapsulation materials/techniques is now comprehensively investigated [125].

To further promote the development of QDSCs, together with performance and stability, the toxicity of the components of these devices should be taken into consideration. For the real and universal application of this new type of solar cells, it is a prerequisite that the whole PV device is environmentally friendly, namely completely "green". For QDSCs, environmental concerns arise from various constituents, such as the high toxicity of Cd-based QD materials and polysulfide redox mediators [124]. Therefore, "green" concepts should be taken into consideration, such as the development of Cd-free I–III–VI group QDs (e.g., $CuInS_2$ and $CuInSe_2$) as alternative light-absorbing materials for QDSCs. From a safety point of view, devices employing aqueous electrolytes are also promising for commercial applications in comparison with solar cells employing conventional toxic-component electrolytes. Regarding the counter electrode, carbon-based materials can be a valuable alternative because of their eco-friendly nature, combined with their low-cost and high catalytic activity. Finally, large-scale manufacturing using high throughput fabrication techniques has been increasingly investigated in the last years by researchers and solar cell developers to achieve real-size PV devices without performance and stability limitations [124].

Perovskite Solar Cells

PSCs belong to the newest technologies of solar cells and have risen to the top of the list of the most promising types of PV devices for commercialization in the near future. Light-to-electricity conversion efficiency in perovskite-based PVs has skyrocketed from 3.8% in 2009 to 25.7% in 2022 in a single junction configuration, and to 31.3% in a perovskite-silicon tandem architecture, setting it the steepest growth ever made in a solar cell technology up to now [3, 126]. This remarkable level of PCE has propelled perovskites to the forefront of cutting-edge optoelectronics research, with materials scientists, chemists, physicists and engineers delving further into their interesting underlying features. With the potential of attaining an even higher performance, along with their low manufacturing cost, these devices have also become attractive for commercialization, with enterprises already investigating industrial plans for entering this technology into the PV market [127].

The term "perovskite" refers to every crystalline material with an ABX_3 structure, originally observed in the mineral calcium titanium oxide ($CaTiO_3$) [4]. Today, almost all of the perovskite materials applied in the PV field are synthesized organic–inorganic hybrids, usually composed of an organic cation such as methylammonium, lead transition metal and a halide anion (see Fig. 4.19). Poly-crystalline perovskites can be easily obtained by simple techniques (e.g., solution-processing) and at low temperatures (i.e., usually between 100 and 150 °C), while their optical bandgap can be tuned from 1.25 to 3.0 eV by simply substituting cation (A, B) or anion (X). Key technical attributes of organic–inorganic perovskites are their low manufacturing cost, long diffusion lengths, low recombination losses and easy tunability

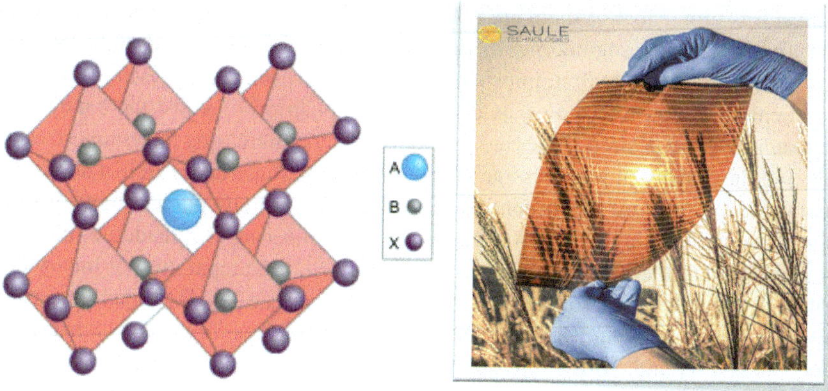

Fig. 4.19 Schematic image of a perovskite crystal (left) and image of a flexible semi-transparent perovskite solar module [132, 133]

of their optical characteristics. Taking advantage of these features, high open-circuit voltage values (>1.1 V) have been attained early in PSCs, typically being the most challenging PV parameter to improve [4].

PSCs design and operation are largely based on DSSCs technology. The architecture of the devices is either developed by mesoscopic or planar structures that may be formed on either a normal or inverted configuration [128]. In mesoscopic designs, the perovskite structures are usually grown as a thin layer that just adequately covers the electron transport material scaffold, while the hole transport material and, subsequently, the back contact are fabricated on top of it. Otherwise, the perovskite precursor solution can be infiltrated into a complete mesoporous device that includes all functional layers of the solar cell (usually developed as a system of electron/insulator/hole transport layers). Regarding the planar architectures, herein, there is a lack of mesoporous scaffolds, while the perovskite structures are layered on top of an electron transport material (normal configuration) or a hole transport material (inverted configuration), followed by the deposition of the opposite charge transport material and finally, the back contact. Perovskite tandems are also possible to be developed, with three main architectures to consider, namely the monolithically integrated, the mechanically stacked and the spectrally split [129]. Today, key challenges that need to be addressed before the entry of PSCs technology in the commercial arena are the refined control of the characteristics of perovskite structures, the poor PSCs stability, the toxicity of some of the component materials of these devices, and finally, the demonstration of the technology on industrial device scales [130, 131].

Under the rapid increase in the PCE to values higher than 25%, PSCs became the game charger in the PV performance race. In order to compete with the electricity cost provided by c-Si solar cells, the refined control of the characteristics of perovskites and, subsequently, the achievement of an even higher and reproducible light-to-electricity conversion efficiency is still considered a primary goal for this technology. Up until today, several groups of strategies have been proposed to meet this aim. Most of them are related to the crystal structure, materials chemistry and processing engineering of perovskite absorbent layer, and others to PSC device physics [134]. At first, improving further the crystal quality of the perovskite absorbent layer and fine-tuning its chemical composition to adjust the bandgap of the material to the desired values are the major targets. Increasing the grains size and reducing the density of pinholes and cracks in the bulk material result in a decreased recombination rate and subsequently enhance the open-circuit photovoltage of solar cells. Amongst the applied approaches, solvent engineering is considered till today the predominant method [135]. Under this approach, the preparation of binary and ternary perovskite precursor solution systems is being recruited for dissolving easier the ionic compounds and forming more stable complexes, both of which greatly affect the quality and morphology of the final crystalline material. The usage of anti-solvent treatments to regulate the nucleation and crystallization of perovskites is also commonly applied to achieve high performance PSCs. Herein, the correct choice of anti-solvent is critical to develop a high-quality perovskite absorbent layer. Polarity, dielectric constant, boiling point and miscibility with water are important attributes of anti-solvent to consider. The volume, the timepoint of dripping and the extrusion rate of the anti-solvent treatment are also important parameters that dominate perovskite film formation, which in turn are highly dependent on the processing conditions [136]. Up until today, a series of anti-solvents have already been employed in PSCs, with special attention to have drawn the green/less hazardous and moisture-resistive anti-solvents that provide an ultrawide processing window. Simultaneously, the introduction of additives in the main solvent or in the anti-solvent is an effective method to improve further the quality of perovskite crystals [137]. These additives can be either organic (i.e., small molecules, fullerenes and polymers) or inorganic (i.e., metal cations, inorganic acids and nanoparticles). The specific role of them depends on their nature. Some additives can trigger the rapid crystallization of perovskites, others can delay the nucleation process, while a combination of these can also be employed to produce more individualized results [135]. To achieve a superior solar cell performance, compositional variation of the perovskite crystal structure is also applied, namely, replacement of commonly used ions with new ones [134]. Today, the composition of perovskite absorbent layers has evolved from simple methylammonium lead triiodide structures to complex mixed composition (multi A cations, mixed B divalent cations and mixed X halides) formamidinium-based perovskite materials. The size and symmetry of the A-site cation greatly affect the bond distance and bond angle between the divalent metal cation and halide anion, which ultimately affects the distribution of the energy band edge states. At the same time, the substitution of the X-site anion significantly changes the optoelectronic characteristics of the material, given that the X-site anion directly contributes to the energy band edge of the perovskite material [134]. Divalent cation variation has also been applied in recent years more and more often as a method to control bandgap and structural characteristics of perovskites [138]. On the other hand, amongst the most effective methods applied today to improve further the performance of PSCs is interfacial engineering. As in other PV technologies, interfacial engineering is an effective route to tune the properties of necessary interfaces and interlayers to overcome interfacial losses by regulating carrier dynamics (i.e., transportation and recombination) and band alignment in the multilayered PSC structure [139]. This involves doping of charge selective layers, the introduction of self-assembled monolayers and buffer layers, and so on. Finally, spectral engineering strategies have been employed more and more in recent years in PSCs, given the development of the materials science field. The plasmon effect, as well as up-conversion and down-conversion strategies, are employed to increase light absorption in the visible spectrum and subsequently enhance the photo-generated current and PCE of PSCs [140].

Considering the golden triangle (performance, cost and stability) to evaluate the commercial feasibility of a solar cell technology, stability is the "Achilles' heel" of PSCs, with the lifetime of these devices to be determined far out of the requirements for a PV product [141]. The topic of stability of PSCs has nowadays seized the paramount attention of the scientific community in the pursuit of attaining their technological maturity for real-world applications. In this run, accelerated ageing and predictive modeling analysis are applied as valuable tools to demonstrate the robustness of PSCs under different environmental conditions in a short testing time [142]. Up until now, several effective modifications of the conventional PSC design have been proposed to deal with the instability of these devices, with their lifetime demonstrating continuous improvement in recent years. Given that the instability of PSCs during their operation mainly stems from the materials intrinsic composition and is self-accelerated by defects in their lattice structure, the focus is greatly put on the perovskite absorbent layer that is determined as the most resistless solar cell material to degradation. Amongst the proposed approaches, component engineering of the ABX_3 structure is considered crucial to develop more operationally stable perovskites [141]. Multi-cation mixed halide perovskites derived after A-site and X-site engineering have been early demonstrated as much more robust crystal structures to degradation compared to the reference methylammonium lead triiodide structures. Additive modifications are also commonly applied to attain defects passivation, morphology improvements and phase stabilization. On the other hand, considering the self-assembly properties of perovskites and the dangling bonds developed on their surface that serve as defect sites where degradation usually starts, surface passivation is commonly employed to improve the operational stability of PSC devices [141]. For modifying perovskite surfaces, organic and inorganic layers, as well as low-dimensional materials, have been commonly employed as surface passivation in PSCs. To increase the stability of PSCs, modification strategies of the other functional layers are also usually employed. These include the modification of both charge selective layers, while interface modifications have also attracted much attention in the past few years to develop permeation barriers to protect perovskites from degradation [141]. Finally, encapsulation is the most effective approach to improve the environmental stability of PSCs. Encapsulation of the multilayered solar cell materials structure not only provides a barrier to prevent the invasion of oxygen and moisture but, more importantly, suppresses the escape of volatile compounds. In this direction, the achievement of a more hermetic enclosure of the core of the PSCs device has been investigated more and more in the last few years [143].

In the last years, the upscaling of PSC technology has attracted considerable attention, accompanied by the rapidly increased PCEs. The aim is to pass from the commonly applied lab-scale spin-coating processing to up-scalable fabrication techniques without deterioration in the performance and stability of the devices [144–146]. To succeed in this, the transition from fully controlled conditions manufacturing to ambient air processing is also considered imperative [147]. Up until today, many research institutes and technical companies have made considerable efforts to advance the technology to industrial scales. Meniscus coating (e.g., blade coating and slot-die coating), screen-printing, spray-coating and inkjet-printing methods are some of the candidates that have already been implemented with great success for the development of this type of devices on a large scale [148]. The optimization of the materials synthesis and fabrication procedures to attain defect-less structures, and, subsequently, highly efficient and stable large-scale PV devices has been increasingly investigated in the last years [149, 150]. On the other hand, several facile strategies have been suggested to deal with the deterioration of solar cell materials under ambient air-processing, especially of the highly moisture-sensitive perovskite absorbent layer [147, 151, 152]. Nowadays, under the great progress achieved in the field, outstanding achievements have already been recorded, with perovskite solar modules efficiency reaching 18% on an 800 cm^2 active area and the first perovskite solar farms to be demonstrated recently [53, 153]. Nevertheless, although the current progress in the scaling up of PSC technology is encouraging, more work should be done to reduce the upscaling performance losses, improve their stability and reduce their manufacturing cost. Additionally, persistent efforts toward the implementation of green chemistry are highly encouraged in PSCs research, not only because sustainable chemistry is ideally inseparable from renewable PV technology but also because the commercialization of PSC practically demands a green technology to reduce its impact on the environment, health and safety [154, 155].

Perovskite-Based Tandem Solar Cells

Currently, as the uptake of PV devices rapidly increases, there are very high expectations from the market makers that the PV sector will continue to be further improved and new solar cell technologies will continue to establish an even lower renewable energy cost. Nevertheless, one of the key basic impediments to this run is that the efficiency of single-junction PV devices is nearing the theoretical limit. On the other hand, multi-junction architectures that are employed in commercialized solar cells to surpass this limit face important obstacles, with the main being the significantly high manufacturing costs, which finally prevent their widespread application. To deal with this challenge, recent developments in new tandem concepts provide a major advance in the field, with perovskite-based tandem solar cells to be the main actor in this run, providing expansive space and abundant opportunity for developing high-efficiency PV devices [156]. In this group of solar cell technologies, perovskite-silicon, perovskite-perovskite, perovskite-CIGS and perovskite-organic tandems are the main architectures that are currently investigated (see Fig. 4.20). These are usually developed based on two typical configurations, namely four-terminal and two-terminal (monolithic).

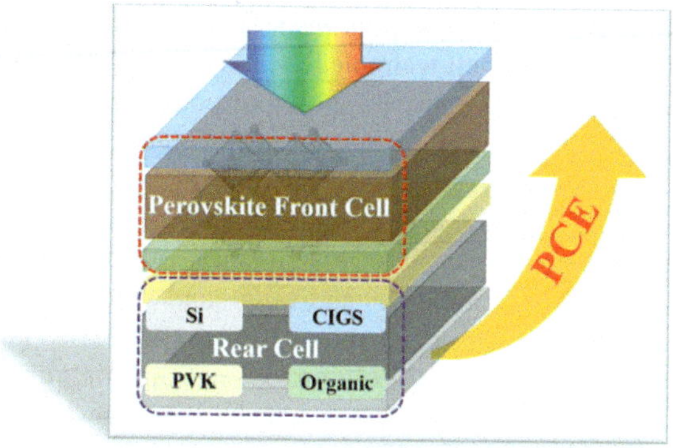

Fig. 4.20 The four major types of perovskite-based tandem solar cells [156]

Among perovskite-based tandems, the perovskite-silicon combination has attracted the greatest attention for building the future of the PV market, mainly stemming from two basic reasons. At first, this is because c-Si solar cells are currently the dominant technology in the PV market, demonstrating high efficiency, excellent stability, mature fabrication and continuously decreasing manufacturing costs. On the other hand, perovskites are considered a wonder PV material and an avenue that provides unique optoelectrical characteristics for high light-to-electricity conversion. Given this, combined with the complementary absorption characteristics to the crystalline silicon that present, perovskite-silicon tandems enable an ultra-wide absorption capability and a great reduction of thermalization losses. In this concept, perovskites are usually employed in front of silicon to absorb high-energy photons, while lower-energy photons are permitted to be absorbed by the c-Si rear cell [156]. Even though this architecture of solar cells has been investigated for only a few years, the efficiency records of these devices have already surpassed the efficiency records of the highest performance single-junction GaAs solar cells, determined on a level of 31% [3].

Owning to the tunable bandgap of perovskite materials, developing all-perovskite tandems is also a hot topic of investigation. In this case, mechanically stacking a wide-bandgap perovskite front cell with a narrow-bandgap perovskite rear cell is among the most straightforward methods. To this aim, replacing Pb with Sn is the most successful method to narrower the bandgap of perovskites [156]. Using this solar cell architecture, flexible solar cells with certified PCE on the order of 25% have been recently demonstrated, outperforming the efficiency of all types of flexible thin-film solar cells developed until today [157]. On the other hand, today, more and more researchers are focused on the development of perovskite-CIGs tandems. Given the several advantages of CIGS technology compared to crystalline silicon, these tandems are considered a valuable approach to achieve ultra-high efficiency and cost-effective all-thin-film solar cells, providing today efficiency records on the order of 25% [3]. Finally, tandem solar cells fabricated by wide bandgap perovskites and narrow bandgap organic materials have experienced significant development in the last few years as an effective way to overcome the thermalization loss of single-junction solar cells and surpass the Shockley-Queisser limit. In contrast to the above-mentioned technologies, these solar cells have attracted the attention of the scientific community because they combine important characteristics such as ultra-low cost and simple manufacturing, lightweight, flexibility and low toxicity. Today, the efficiency records of perovskite-organic tandems are determined on the order of 20%. Yet, based on the semi-empirical device models, it is predicted that a practical efficiency of over 30% is attainable with the rapid development of low-band-gap organic semiconductors and wide-band-gap perovskite materials [158].

Although the development of perovskite-based tandem architectures stands out as a valuable strategy to achieve low-cost and high-efficiency PV devices, several challenges appear in the design and characteristics of this kind of solar cells that should be tackled to attain their widespread development and commercialization [156]. At first, key obstacles are related to the device structure, both for four-terminal and two-terminal configurations. Four-terminal tandems usually suffer from parasitic absorption and reflection of light from the additional transparent electrodes and interfaces that are composed, which ultimately decrease their PCE. On the other hand, two-terminal configurations require very good current matching between the two sub-cells to attain a low energy loss, which is quite difficult to achieve in practice. Besides, in their manufacturing processing, the contamination of under-layered sub-cell by impurities and halide ion migration arising from the perovskite upper layer is easy to happen and should be taken care of. Regarding the efficiency of these devices, a key challenge for further improvements is the high energy loss attributed to the high nonradiative recombination at the grain boundaries and interfaces of wide-bandgap perovskite that are usually applied as light active materials in the front sub-cells. Inferior energy alignment is also commonly observed between perovskites and charge-transport layers, that also decreases solar cells performance. On the other hand, stability issues appear in this kind of devices, mainly attributed to the easy deterioration of the characteristics of perovskite layers when meeting light, oxygen, moisture and heating, while the degradation of the devices efficiency is also common when hydrophilic compounds are employed. Large-area fabrication is also a key milestone that should be demonstrated to confirm the industrial feasibility of perovskite-based tandems since most of the high-performance devices are still at the laboratory level. This, of course, should be accompanied by a great reduction of manufacturing costs in order to enable this technology to provide a competitive light-to-electricity conversion compared to the already established solar cell technologies. Finally, lead toxicity is an important issue that needs to be considered. Although lead-free perovskites have already been developed as alternatives, these materials suffer from inferior performance and stability compared to conventional lead-based perovskites. Nevertheless, based on the rapid progress achieved in recent years in the field, it is believed that the perovskite-based tandem solar cells will continue their great development, and possibly, it will be the next generation of solar cells to dominate the PV market.

4.6 Life Cycle Assessment Data for Solar Cell Technologies Toward the Design of Greener Agrivoltaic Energy Systems

The methodological framework of life cycle assessment (LCA) is widely employed in the PV field to estimate and assess the environmental impacts attributable to the life cycle of a PV system. LCA of PV systems is very important since the quantification of the technological, environmental, economic and social aspects of PVs have a key role in reducing the life cycle impacts of these systems toward more sustainable options. This kind of evaluation is of utmost importance for the strategic development of novel agrivoltaic technologies toward the establishment of a real sustainable agriculture.

The life cycle of a PV system begins with the raw materials extraction (cradle) and ends with PV components disposal or recycling (grave) (see Fig. 4.21). Raw materials, energy supply, panels manufacturing, mounting system, cables, inverters and every additional component needed for the fabrication of a PV system is included in the product stage. Transportation, construction and installation are included in the construction stage. Auxiliary electricity demand, panels cleaning and maintenance, as well as repair and replacement (if needed), belong to the usage state. Finally, recycling and reuse, waste processing and transportation regard the end-of-life stage. All these stages define the system boundary of the life cycle of a PV system.

In general, the LCA methodology consists of four phases, according to the standardized framework series of ISO 14040 and ISO 14044. These phases involve (a) the goal and scope definition, where the objective, system boundaries, functional unit and assumptions are determined; (b) the life cycle inventory (LCI), which regards

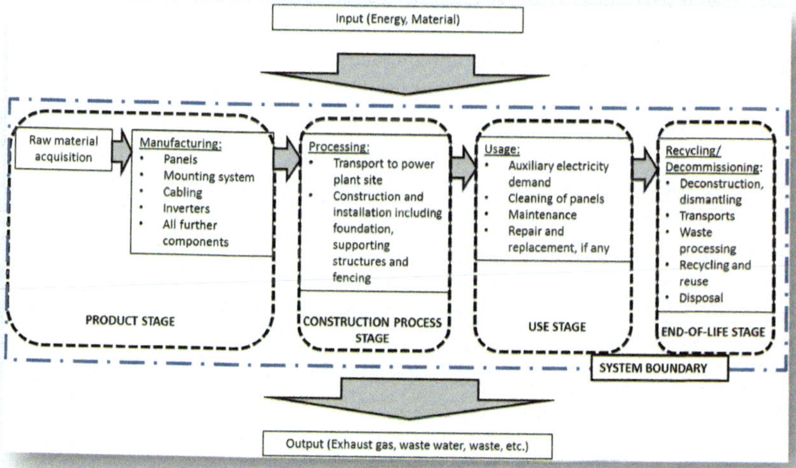

Fig. 4.21 The system boundary of the life cycle of a photovoltaic system [159]

the data collection in terms of the inputs, intermediate processes and outputs; (c) the life cycle impact assessment (LCIA), where the potential environmental impacts of the system are defined and finally, (d) the interpretation, where LCI and LCIA results are summarized, the critical points are identified and analyzed, and conclusions and recommendations for improvements are made [159].

In the LCA of a PV system, the following parameters should be presented: (1) aim of the study and functional unit; (2) production assumptions for the key input materials; (3) location of the installation; (4) annual irradiation; (5) solar cell technology; (6) type of system (e.g., roof-top or ground-mounted, fixed-tilted or with a tracker); (7) module efficiency and degradation rate; (8) lifetime of PVs and BOS; (9) expected annual electricity production; (10) timeframe of data; (11) life cycle stages included; (12) LCA approach and software tool used; (13) LCI databases used; (14) place of production modeled; and (15) electricity source. These parameters are noteworthy to be reported, not only to increase the transparency of the analysis but also because they can significantly affect the LCA findings when they vary [160].

Herein, a short review on the LCA of the different solar cell technologies is conducted, aiming to provide a solid foundation for future LCA research on the next-generation agrivoltaic energy systems. The review is mainly focused on presenting data on the metrics of energy payback time (EPBT) and greenhouse gas (GHG) emissions that are the most commonly used indicators in PV LCA studies. Toxicity, acidification and resource depletion impacts are also discussed for a more comprehensive point of view on the topic.

Crystalline Silicon Solar Cells

The production processing of mc-Si and sc-Si solar cells is quite similar, although some differences in their environmental impacts appear due to differences in some of the steps of their manufacturing [161]. Their development begins with the production of SiO_2 from the mining of quartz sand that is subsequently reduced to metallurgical-grade silicon. Then, silicon is purified into solar-grade silicon (>99.99% purity) through various methods, where working temperatures above 1100 °C are applied, leading to the melting of silicon and the production of crystalline silicon ingots. After this step, the ingots are sliced into wafers, usually of 156×156 mm^2 size, and subsequently, solar cells are produced by etching, doping, screen printing, coating and checking [159].

Up until now, numerous LCA studies have already been conducted on mc-Si and sc-Si technologies, the majority of which were carried out after 2000. Taking into account that these PV technologies present a rapid and continuous development, the input of data derived from studies held in the last few years is quite important to be considered in order to achieve validity. Based on recent publications on c-Si solar cells, mc-Si and sc-Si technologies present a similar range of values for GHG emissions and EPBT, with most of them being determined on the order of 20–50 gCo_{2eq}/kWh and 1–2.5 years, respectively. Of course, these values vary depending on the assumptions made [162]. Most of the environmental burdens come from the production stage of PVs. The PV manufacturing process occupies about 85% of the total energy usage and total GHG emissions. From this, the production process of solar-grade silicon accounts for more than 35% of the total energy consumption and GHG emissions [163]. Efficiency is also a critical parameter that affects the environmental impacts of PVs. Studies have shown that more than 25% of the overall environmental burden could be efficiently reduced by improving energy efficiency, choosing secondary aluminum for c-Si production and reducing c-silicon wafer consumption for PV production [164]. The location of installation is also critical. The environmental impact of PVs installed in locations with high solar irradiation is usually much lower compared to the corresponding cases of locations with low solar irradiation [165]. Among the c-Si technologies, mc-Si solar cells present a lower environmental impact compared to the sc-Si ones [166]. Some recent studies also demonstrate that the advanced c-Si solar cell technologies, such as PERC, present more environmentally friendly characteristics (always in terms of GHG emissions and EBPT) compared to the reference Al-BSF technology [167, 168]. Besides, the replacement of solar-grade silicon feedstocks with upgraded metallurgical-grade silicon ones can lead to a further decrease of over 20% in climate change emissions and over 25% in EPBT [169]. Regarding the environmental performance of c-Si PVs on the module level, this highly depends on the module technology. Studies have shown that with a frameless double-glass structure, a reduction in the values of GHG emissions and EPBT can be achieved compared to the conventional glass-backsheet structure [170]. Finally, taking into account the recent strategies for recycling end-of-life c-Si PV modules, studies have shown that the GHG emissions can be reduced by 5 times compared to the cases of landfilling [171]. On the other hand, the use of the recycled silicon material in c-Si PV modules production can lead to a reduction of GHG emissions at the level of approximately 40%, mainly deriving from the reduction of energy consumption during manufacturing [172].

Commercialized Thin-Film Solar Cells

Considering that a-Si:H, CIGS and CdTe solar cell technologies are developed to reduce material and energy consumption compared to the c-Si solar cells, LCA studies on the former technologies present an increasing interest in the last few years to estimate and assess the environmental impacts attributed to their life cycle. This is very important, especially for the cases of new applications that support, where requirements can be significantly higher compared to the corresponding ones of conventional terrestrial applications.

Starting with the a-Si:H solar cells that have the longest time in the commercial arena, many studies demonstrate that they can yield a greater net environmental benefit compared to crystalline silicon technology [173]. In a-Si solar cells, the active silicon-based material is developed in thin-film layers of thicknesses less than 500 nm, using significantly lower raw material quantities for manufacturing compared to c-Si solar cells and non-toxic dopants. Besides, the processing temperatures usually do not exceed even the 100 °C. Thus, a significantly lower amount of energy is consumed during the production stage. According to LCA studies on a-Si solar cells, the GHG emissions of this technology are usually determined below 40 gCo_{2eq}/kWh and the EPBT on the order of 2 years [162]. The reason behind the small differences in the reported values compared to the corresponding ones of the energy-intensive c-Si solar cells is that the a-Si solar cell technology usually leads to a much lower PCE and higher BOS requirements. Of course, these numbers vary greatly, depending on the assumptions made. For instance, encapsulation (during the production stage) and roof integration (during the construction stage) appear as the main contributors to their global warming potential [174]. Thus, it is critical to pay attention to these steps to decrease GHG emissions. Other studies show that the EPBT of these solar cells can be almost halved when a frameless design is employed compared to that of framed a-Si PV systems, and so on [175].

CIS and CIGS solar cells, on the other hand, are also promising alternatives to c-Si solar cells to reduce the environmental impacts of light-to-electricity conversion. Taking into account that they are developed in thin-films, they have all the benefits of thin-film solar cells, such as the low materials and manufacturing energy demands, which, when combined with the non-toxicity of chalcogenides, lead to low GHG emissions. Simultaneously, by considering their satisfactory efficiency, which does not differ intensively from that of c-Si solar cells, the EPBT is also at a competitive level [175]. According to LCA studies on CIS and CIGS solar cells, the GHG emissions of these technologies are usually determined below 40 gCo_{2eq}/kWh and the EPBT around 1.5 years [162].

Finally, the LCA of CdTe PV systems has gained increasing attention in the last decade since CdTe is the technology with the highest market share amongst commercialized thin-film solar cells. LCA studies on CdTe solar cells usually demonstrate GHG emissions below 30 gCo_{2eq}/kWh and an EPBT on the order of 1–1.5 years [162]. A great part of the total global warming potential of this technology usually arises from the life cycle GHG emissions of the panels, while a much lower part is attributed to the BOS. The electricity used to produce CdTe PV panels is the main contributor to the global warming potential of the panels. Nevertheless, LCA studies have shown that with a further increase in the PCE of the solar cells, a significant decrease in their GHG emissions can be achieved, giving potential for further improving the environmental profile of CdTe PVs [176]. Today, CdTe solar cell technology is set among the technologies with the lowest LCA impact, with the only significant emission that exists to be the discharge of toxic cadmium, although it is considered that this emission is relatively low compared to the life cycle emissions of other PV plants [175].

Single-Junction Gallium Arsenide and III–V Multi-junction Solar Cells

GaAs solar cells and III–V MJ PVs have historically demonstrated the highest PCEs compared to any other PV technology developed up until now, but are very expensive as well. Thus, their development for terrestrial application is very limited and the existing LCA studies on the aforementioned technologies are very few, and most of them unavailable to the broad public (scientific community and external stakeholders). Among the few available ones is that conducted by Mohr et al., who did a life cycle assessment comparing the limited industrial production scale of 100 kW_p per year for GaAs and GaInP/GaAs tandem solar cells, and 10 MW_p per year for an mc-Si technology, considering different environmental indicators [177]. Their results showed that the overall environmental impact of the production of the III–V modules is larger than the corresponding of the common multi-Si module production, with the epitaxy process being the main contributor to the energy consumption. In another investigation carried out on III–V MJ high-concentration PV system, the results showed that the cumulative energy demand for this kind of systems is an order of magnitude higher than the corresponding of c-Si PV technologies, setting further optimizations as necessary to achieve competitiveness in the widespread commercial arena of terrestrial PV applications [178].

Emerging Solar Cell Technologies

The demand for safer, greener and more sustainable PVs has evolved rapidly in the past years. Emerging solar cell technologies are continuously developed to achieve a smaller environmental footprint during their life cycle and widespread commercialization, aiming for new and established applications. LCA studies on these technologies are increasingly attracting the attention of the scientific community and industry as their performance, stability and cost come to a competitive level compared to the established market. More and more LCI data are provided following this rapid evolution to keep the validity of the LCA results.

Amongst the innovative technologies developed, DSSCs stand out as economically and environmentally viable alternatives to conventional PV devices. The industrial production of DSSCs is considered very similar to those employed for the production of other thin-film solar module technologies. The main steps characterizing the production process of a DSSC module in the Grätzel prototype standard configuration can be summarized as follows: glass substrate coating, laser etching, deposition of active layers, sintering, staining, substrates matching and electrolyte filling. According to LCA studies on this technology, the cumulative energy demand, global warming potential and EPBT values are comparable and sometimes better than the corresponding ones of inorganic thin-film devices. In fact, an energy saving on the order of 50% and a reduction of CO_2 emissions of 60% are achievable by employing dye-sensitized solar modules for PV installation. This turns out in an EPBT saving ranging from a minimum of 8 to a maximum of 14 months with respect to other thin film panels [179].

OPVs, on the other hand, are often proposed as a solution to achieve low cost and low environmental impact in light-to-electricity conversion. OPVs are mainly composed of low-cost and non-toxic materials such as the PET substrate and small amounts of organic semiconductors and organic substances. Today, LCA studies are carried out on organic solar cells to evaluate a priori the environmental impact of large-scale production and identify pathways toward their sustainable development. According to LCA studies on various types of OPVs, the results showed that the EPBT of this kind of solar cell technology could be determined below 1 year [180]. This, of course, depends on the assumptions made, such as the solar cell design and the fabrication technique applied. Obviously, there are uncertainties surrounding the OPV panels environmental assessment, such as the efficiency and the lifetime of the devices at the industrial scale. The LCA studies demonstrate that OPVs have the greatest potential to improve the environmental performance of the PV industry when they are used in short-term applications involving very simple (i.e., lack of ancillary components) and comparatively low-mass devices (i.e., low resource demanding) [181].

Kesterite-based solar cells, such as CZTS, are also one of the most promising green alternatives for light-to-electricity conversion since they are made from earth-abundant and low-toxicity materials. In order to assess the potential of CZTS PVs in the future global market, several studies are nowadays conducted to investigate their sustainability by assessing the possible environmental impacts of their large-scale production. LCA studies on CZTS solar cells considering vacuum and non-vacuum fabrication techniques have already been conducted, using information from laboratory-scale processes as well as from large-scale estimations [182]. In most of them, CIGS solar cells are selected as the commercial reference. The analyses suggested that the absorber layers are the most demanding manufacturing stage, contributing the most to the majority of the impact categories assessed, although CZTS films fabricated with either vacuum or non-vacuum techniques exhibit a significant environmental potential benefit over the CIGS films. Among the vacuum and non-vacuum techniques, the energy consumption for processing CZTS films is considerably lower in the latter case compared to that required for the former ones [182]. However, because the results regarding the environmental performance of these emerging solar cells are sensitive to changes in efficiency, to achieve a clear superiority, higher PCEs should be achieved. The replacement of materials and chemicals with environmentally friendlier alternatives can also improve the environmental performance of these solar cells. Recycling cell materials at the end of life could also contribute greatly to the reduction of the associated impact by extending the lifetime of the materials and keeping them in the cycle of use for longer. However, recycling can be problematic as the quantity of materials that can be recovered at the end of life from the solar cells is usually low and economically unattractive [183].

QDSCs were introduced in the past few years to achieve efficient conversion of solar energy into electric power, as a promising and with unique characteristics alternative PV technology. Many techniques have been employed to prepare QD sensitizers coated on mesoporous scaffolds, either by in situ or ex situ fabrication, many of them being straightforward, energy advantageous and up-scalable. Up until today, there are some comprehensive LCA studies demonstrating the environmental performance of these devices. From these, the values of their GHG emissions and EPBT are determined on the level of 3 gCO_{2eq}/kWh and 1.5 years, respectively. Others demonstrate that GHG emissions and EPBT are even lower than CIS technology [159]. Of course, there is a need for further studies and updated LCI data to attain validity.

Finally, PSCs have moved to the forefront of emerging thin-film solar cell research in just a decade by demonstrating very promising efficiency records. With the continuous record-breaking PCEs in the past few years, which now exceed even the records of mc-Si solar cells, PSCs are considered a great challenger to the already commercial PV technologies. Given that their stability is continuously improved, their environmental sustainability is today a topic of attention from both the scientific community and industry, with a number of LCA studies on perovskite technology appearing.

According to them, the manufacturing processing of perovskite solar modules can cause up to 30% lower impacts compared to that of sc-Si solar cells, although the total environmental impacts coming from their lifetime are calculated higher than the latter technology, mainly arising from their much lower stability [184]. The main energy consumption of the life cycle of PSCs originates from their production process, with the total energy requirements being higher in the case of n-i-p architecture than the p-i-n architecture [185]. Of course, the energy requirements depend highly on the fabrication method of the devices, with the inkjet-printing giving considerable environmental advantages compared to the commonly applied techniques [186]. The major environmental hotspots of PSC technology come from the use of gold and the energy-intensive thermal evaporation, while lower (on the order of 25%) is their total environmental impact when a hole-transport-material-free design is employed [184, 187]. On the other hand, PSCs can provide the shortest EPBT among existing PV technologies, even from c-Si solar cells and CdTe solar cells, with the values to be determined on the order of 1 year. Recycling strategies are intended to lead to a high decrease in both GHG emissions and EPBT, which can be reduced even by 70%. In this way, values on the order of 15 gCO_{2eq}/kWh and 0.1 years can be attained, which outcompete all other rivals, including the market-leading silicon [188].

4.7 Outlook

Predicting the future development of any solar cell technology is inherently risky. While c-Si solar cells now dominate the PVs industry, alternative technologies are fast advancing. The future of light-to-electricity conversion may be a refined version of today's commercial solar cells or a totally new solar cell technology. In the face of unknown technological advancements and uncertain economic pressures, it is best to avoid banking on any certain solar cell technology. Instead, all technologies should be continuously reviewed objectively through the perspective of application-driven performance metrics. These indicators guide a solar cell designer in the direction of technical themes such as higher efficiency and stability, decreased manufacturing complexity and cost, and a broad application range. Taking into account the aforementioned, the development of the next generation agrivoltaics will depend on the course of development of each of the aforementioned solar cell technology, the adoption of each or their combination will provide customized solutions for each strategic plan to meet sustainability in the agricultural sector.

References

1. Fraas LM (2014) Low-cost solar electric power, first. Springer, Bellevue
2. Chapin DM, Fuller CS, Pearson GL (1954) A new silicon p-n junction photocell for converting solar radiation into electrical power. J Appl Phys 25:676–677
3. (2023) National renewable energy laboratory: photovoltaic research. https://www.nrel.gov/pv/. Accessed 06 Dec 2023

4. Jean J, Brown PR, Jaffe RL et al (2015) Pathways for solar photovoltaics. Energy Environ Sci 8:1200–1219
5. (2007) Bright outlook for solar cells. Phys. World. https://physicsworld.com/a/bright-outlook-for-solar-cells/. Accessed 5 Oct 2022
6. Green MA (1994) Silicon solar cells: the ultimate photovoltaic solution? Prog Photovoltaics Res Appl 2:87–94
7. Letcher T (2022) Comprehensive renewable energy, second. Elsevier
8. Fthenakis V, Lynn P (2018) Electricity from sunlight: photovoltaic-systems integration and sustainability, second. Wiley
9. Zhou Y (2013) Eco- and renewable energy materials, first. Springer
10. Solarlink (2022) N-type vs p-type panels—which should you choose? https://solarlinkaustralia.com.au/n-type-vs-p-type-panels-which-should-you-choose/. Accessed 6 Oct 2022
11. Pickerel K (2018) The difference between n-type and p-type solar cells. https://www.solarpowerworldonline.com/2018/07/the-difference-between-n-type-and-p-type-solar-cells/. Accessed 6 Oct 2022
12. Helmoholtz Zentrum Berlin (2022) Silicon heterojunction solar cells. https://www.helmholtz-berlin.de/projects/pvcomb/forschen/silizium/pvcomb---hjt-cells_en.html. Accessed 6 Oct 2022
13. Liu J, Yao Y, Xiao S, Gu X (2018) Review of status developments of high-efficiency crystalline silicon solar cells. J Phys D Appl Phys 51:123001
14. Kowalczewski P, Andreani LC (2015) Towards the efficiency limits of silicon solar cells: how thin is too thin? Sol Energy Mater Sol Cells 143:260–268
15. Imec (2019) Imec and Jolywood achieve a record of 23.2 percent with bifacial n-PERT solar cells. https://www.imec-int.com/en/articles/imec-and-jolywood-achieve-a-record-of-23-2-percent-with-bifacial-n-pert-solar-cells. Accessed 8 Aug 2023
16. Green MA, Blakers AW, Zhao J et al (1990) Characterization of 23-percent efficient silicon solar cells. IEEE Trans Electron Devices 37:331–336
17. Chan KS (2019) What is a PERT solar cell? https://www.kschan.com/what-is-a-pert-solar-cell/. Accessed 6 Oct 2022
18. Feldmann F, Bivour M, Reichel C et al (2014) Passivated rear contacts for high-efficiency n-type Si solar cells providing high interface passivation quality and excellent transport characteristics. Sol Energy Mater Sol Cells 120:270–274
19. Solar Era (2022) IBC vs. PERC: What's the Best Type of Solar Panel. https://www.thesolarera.com/post/ibc-vs-perc-what-s-the-best-type-of-solar-panel. Accessed 1 Aug 2023
20. Nakamura J, Asano N, Hieda T et al (2014) Development of heterojunction back contact Si solar cells. IEEE J Photovoltaics 4:1491–1495
21. Yoshikawa K, Yoshida W, Irie T et al (2017) Exceeding conversion efficiency of 26% by heterojunction interdigitated back contact solar cell with thin film Si technology. Sol Energy Mater Sol Cells 173:37–42
22. VDMA Photovoltaic Equipment (2022) International technology roadmap for photovoltaic (ITRPV). Frankfurt
23. Kopecek R, Libal J (2021) Future industrial solar PV technologies: champion cell announcements versus industrial reality. https://www.pv-tech.org/future-industrial-solar-pv-technologies-champion-cell-announcements-versus-industrial-reality/. Accessed 6 Oct 2022
24. McCann MJ, Catchpole KR, Weber KJ, Blakers AW (2001) A review of thin-film crystalline silicon for solar cell applications. Part 1: native substrates. Sol Energy Mater Sol Cells 68:135–171
25. VDMA (2018) International technology roadmap for photovoltaic. Frankfurt
26. Ibn-Mohammed T, Koh SCL, Reaney IM et al (2017) Perovskite solar cells: an integrated hybrid lifecycle assessment and review in comparison with other photovoltaic technologies. Renew Sustain Energy Rev 80:1321–1344
27. Stuckelberger M, Biron R, Wyrsch N et al (2017) Review: progress in solar cells from hydrogenated amorphous silicon. Renew Sustain Energy Rev 76:1497–1523
28. Fraunhofer (2018) Photovoltaics report. Freiburg

29. Sogar T (2006) Nanostructured materials for solar energy conversion, first. Elsevier, Nagoya
30. Ramanujam J, Bishop DM, Todorov TK et al (2020) Flexible CIGS, CdTe and a-Si: H based thin film solar cells: a review. Prog Mater Sci 110:100619
31. Wieder S (1999) Amorphous silicon solar cells comparison of p-i-n and n-i-p structures with zinc-oxide front contact. Forschungszentrums Jülich
32. Sreejith S, Ajayan J, Kollem S, Sivasankari B (2022) A comprehensive review on thin film amorphous silicon solar cells. SILICON 14:8277–8293
33. Yang X-Y (2019) Photoenergy and thin film materials, first. Wiley Online Library, Hoboken, New Jersey
34. Mufti N, Amrillah T, Taufiq A et al (2020) Review of CIGS-based solar cells manufacturing by structural engineering. Sol Energy 207:1146–1157
35. Midsummer (2021) Midsummer BOLD. https://midsummer.se/en/midsummer-bold/. Accessed 4 Aug 2023
36. Letcher TM, Fthenakis VM (2018) A comprehensive guide to solar energy systems, first. Academic Press, New York
37. Romeo A, Artegiani E, Menossi D (2018) Low substrate temperature CdTe solar cells: a review. Sol Energy 175:9–15
38. U.S. Department of Enegy Cadmium Telluride. https://www.energy.gov/eere/solar/cadmium-telluride. Accessed 11 Oct 2022
39. Burst JM, Duenow JN, Albin DS et al (2016) CdTe solar cells with open-circuit voltage breaking the 1 V barrier. Nat Energy 1:16015
40. Zhao Y, Boccard M, Liu S et al (2016) Monocrystalline CdTe solar cells with open-circuit voltage over 1 V and efficiency of 17%. Nat Energy 1:16067
41. Wikipedia Thin-film solar cell. https://en.wikipedia.org/wiki/Thin-film_solar_cell. Accessed 11 Oct 2022
42. Glava Energy Center Flexible thin-film solar cells application in real life. https://glavaenergycenter.se/en/. Accessed 14 Aug 2023
43. a2 solar: advanced & automotive. https://a2-solar.com/en/automotive-pv/car/. Accessed 11 Oct 2022
44. Sunslice Fusion flex 6 watts - portable solar panel. https://sunslice-solar.com/products/flexible-solar-charger-for-smartphone-6w. Accessed 8 Jul 2023
45. Cotal H, Fetzer C, Boisvert J et al (2009) III–V multijunction solar cells for concentrating photovoltaics. Energy Environ Sci 2:174–192
46. (2017) NREL licenses inverted metamorphic multi-junction solar cell technology. https://www.solarpowerworldonline.com/2017/10/nrel-licenses-inverted-metamorphic-multi-junction-solar-cell-technology/. Accessed 11 Oct 2022
47. Henry CH (1980) Limiting efficiencies of ideal single and multiple energy gap terrestrial solar cells. J Appl Phys 51:4494–4500
48. Miles RW, Hynes KM, Forbes I (2005) Photovoltaic solar cells: an overview of state-of-the-art cell development and environmental issues. Prog Cryst Growth Charact Mater 51:1–42
49. Wikipedia Concentrator photovoltaics. https://en.wikipedia.org/wiki/Concentrator_photovoltaics#:~:text=Concentrator photovoltaics (CPV) (also, junction (MJ) solar cells. Accessed 12 Oct 2022
50. Horowitz KAW, Remo T, Smith B, Ptak A (2018) A techno-economic analysis and cost reduction roadmap for III–V solar cells. Denver
51. O'Regan B, Grätzel M (1991) A low-cost, high-efficiency solar cell based on dye-sensitized colloidal TiO_2 films. Nature 353:737–740
52. Kakiage K, Aoyama Y, Yano T et al (2015) Highly-efficient dye-sensitized solar cells with collaborative sensitization by silyl-anchor and carboxy-anchor dyes. Chem Commun 51:15894–15897
53. Green MA, Dunlop ED, Hohl-Ebinger J et al (2022) Solar cell efficiency tables (Version 60). Prog Photovoltaics Res Appl 30:687–701
54. Baxter JB (2012) Commercialization of dye sensitized solar cells: present status and future research needs to improve efficiency, stability, and manufacturing. J Vac Sci Technol A Vacuum, Surfaces, Film 30:020801

55. Hagfeldt A, Boschloo G, Sun L et al (2010) Dye-sensitized solar cells. Chem Rev 110:6595–6663
56. Mozaffari S, Nateghi MR, Zarandi MB (2017) An overview of the challenges in the commercialization of dye sensitized solar cells. Renew Sustain Energy Rev 71:675–686
57. Solaronix. www.solaronix.com/solarcells/developments/. Accessed 12 Oct 2022
58. Brite Solar. https://www.britesolar.com/latest-greenhouse-coverings-technology/. Accessed 12 Oct 2022
59. xsreviews (2016) GCELL solar backpack & solar keyboard folio review: from wales with love. https://xsreviews.co.uk/reviews/gcell-solar-backpack-solar-keyboard-folio-review-from-wales-with-love/. Accessed 10 Aug 2023
60. Uchida S No Title. http://dsc.rcast.u-tokyo.ac.jp/~dsc/modules/SONY/DSC01596s.jpg. Accessed 10 Jul 2023
61. Shockley W, Queisser HJ (1961) Detailed balance limit of efficiency of p-n junction solar cells. J Appl Phys 32:510–519
62. Hardin BE, Snaith HJ, McGehee MD (2012) The renaissance of dye-sensitized solar cells. Nat Photonics 6:162–169
63. Fan K, Yu J, Ho W (2017) Improving photoanodes to obtain highly efficient dye-sensitized solar cells: a brief review. Mater Horizs 4:319–344
64. Chalkias DA (2019) Fabrication and characterization of novel hybrid nanocomposites with application in solar cells. University of Patras
65. Wu J, Lan Z, Lin J et al (2015) Electrolytes in dye-sensitized solar cells. Chem Rev 115:2136–2173
66. Thomas S, Deepak TG, Anjusree GS et al (2014) A review on counter electrode materials in dye-sensitized solar cells. J Mater Chem A 2:4474–4490
67. Wu J, Lan Z, Lin J et al (2017) Counter electrodes in dye-sensitized solar cells. Chem Soc Rev 46:5975–6023
68. Chalkias DA, Loizos DD, Papanicolaou GC (2020) Evaluation and prediction of dye-sensitized solar cells stability under different accelerated ageing conditions. Sol Energy 207:841–850
69. Chalkias DA, Verykokkos NE, Kollia E et al (2021) High-efficiency quasi-solid state dye-sensitized solar cells using a polymer blend electrolyte with "polymer-in-salt" conduction characteristics. Sol Energy 222:35–47
70. Bella F, Verna A, Gerbaldi C (2018) Patterning dye-sensitized solar cell photoanodes through a polymeric approach: a perspective. Mater Sci Semicond Process 73:92–98
71. Chalkias DA, Giannopoulos DI, Kollia E et al (2018) Preparation of polyvinylpyrrolidone-based polymer electrolytes and their application by in-situ gelation in dye-sensitized solar cells. Electrochim Acta 271:632–640
72. Figgemeier E, Hagfeldt A (2004) Are dye-sensitized nano-structured solar cells stable? An overview of device testing and component analyses. Int J Photoenergy 6:127–140
73. Luo F, Wang L, Ma B, Qiu Y (2008) Post-modification using aluminum isopropoxide after dye-sensitization for improved performance and stability of quasi-solid-state solar cells. J Photochem Photobiol A Chem 197:375–381
74. Pathak SK, Abate A, Ruckdeschel P et al (2014) Performance and stability enhancement of dye-sensitized and perovskite solar cells by Al doping of TiO_2. Adv Funct Mater 24:6046–6055
75. Murakami TN, Yoshida E, Koumura N (2014) Carbazole dye with phosphonic acid anchoring groups for long-term heat stability of dye-sensitized solar cells. Electrochim Acta 131:174–183
76. Asghar MI, Miettunen K, Halme J et al (2010) Review of stability for advanced dye solar cells. Energy Environ Sci 3:418–426
77. Li H, Tang Q, Meng Y et al (2016) Dissolution-resistant platinum alloy counter electrodes for stable dye-sensitized solar cells. Electrochim Acta 190:409–418
78. Feng C, Jin Z, Zhang M, Wang Z-S (2018) Nanocomposites of nickel selenide supported on cube-shaped lidless graphitic boxes as efficient counter electrodes for quasi-solid-state dye-sensitized solar cells. Electrochim Acta 281:237–245

79. Gao J, Yang Y, Zhang Z et al (2016) Bifacial quasi-solid-state dye-sensitized solar cells with poly (vinyl pyrrolidone)/polyaniline transparent counter electrode. Nano Energy 26:123–130
80. Freitag M, Teuscher J, Saygili Y et al (2017) Dye-sensitized solar cells for efficient power generation under ambient lighting. Nat Photonics 11:372–378
81. Xie K, Guo M, Liu X, Huang H (2015) Enhanced efficiencies in thin and semi-transparent dye-sensitized solar cells under low photon flux conditions using TiO_2 nanotube photonic crystal. J Power Sources 293:170–177
82. Hashmi G, Miettunen K, Peltola T et al (2011) Review of materials and manufacturing options for large area flexible dye solar cells. Renew Sustain Energy Rev 15:3717–3732
83. Chalkias DA, Laios AI, Petala A, Papanicolaou GC (2018) Evaluation of the limiting factors affecting large-sized, flexible, platinum-free dye-sensitized solar cells performance: a combined experimental and equivalent circuit analysis. J Mater Sci Mater Electron 29:9621–9634
84. Grifoni F, Bonomo M, Naim W et al (2021) Toward sustainable, colorless, and transparent photovoltaics: state of the art and perspectives for the development of selective near-infrared dye-sensitized solar cells. Adv Energy Mater 11:2101598
85. Hashmi SG, Özkan M, Halme J et al (2016) Dye-sensitized solar cells with inkjet-printed dyes. Energy Environ Sci 9:2453–2462
86. Rui Y, Wang Y, Zhang Q et al (2016) In-situ construction of three-dimensional titania network on Ti foil toward enhanced performance of flexible dye-sensitized solar cells. Appl Surf Sci 380:210–217
87. Yun MJ, Cha SI, Seo SH, Lee DY (2014) Highly flexible dye-sensitized solar cells produced by sewing textile electrodes on cloth. Sci Rep 4:5322
88. Liang J, Zhang G, Yang Y, Zhang J (2014) Highly ordered hierarchical TiO_2 nanotube arrays for flexible fiber-type dye-sensitized solar cells. J Mater Chem A 2:19841–19847
89. Obaidullah M, Esat V, Sabah C (2018) Single- and multi-walled carbon nanotubes for solar cell applications. Int J Mod Phys B 32:1830007
90. Liu W, Zhang H, Wang H et al (2017) Titanium mesh supported TiO_2 nanowire arrays/upconversion luminescence Er^{3+}-Yb^{3+} codoped TiO_2 nanoparticles novel composites for flexible dye-sensitized solar cells. Appl Surf Sci 422:304–315
91. Park JH, Jun Y, Yun H-G et al (2008) Fabrication of an efficient dye-sensitized solar cell with stainless steel substrate. J Electrochem Soc 155:F145–F149
92. Lee C-P, Lin RY-Y, Lin L-Y et al (2015) Recent progress in organic sensitizers for dye-sensitized solar cells. RSC Adv 5:23810–23825
93. Hao F, Lin H, Liu Y et al (2011) Evidence for enhancing charge collection efficiency with an alternative cost-effective binary ionic liquids electrolyte based dye-sensitized solar cells. Electrochim Acta 56:5605–5610
94. Karthika P, Ganesan S, Arthanareeswari M (2018) Low-cost synthesized organic compounds in solvent free quasi-solid state polyethyleneimine, polyethylene glycol based polymer electrolyte for dye-sensitized solar cells with high photovoltaic conversion efficiencies. Sol Energy 160:225–250
95. Ganesan S, Mathew V, Rathika R et al (2018) Low-cost tetra ethylene glycol derivatives in polymer blend electrolytes for dye-sensitized solar cells with high photovoltaic conversion efficiencies. Mater Sci Eng B 229:37–43
96. Xu B, Bi D, Hua Y et al (2016) A low-cost spiro[fluorene-9,9′-xanthene]-based hole transport material for highly efficient solid-state dye-sensitized solar cells and perovskite solar cells. Energy Environ Sci 9:873–877
97. Makris T, Mourtzikou A, Rapsomanikis A, Stathatos E (2017) Inkjet printable processes for dye-sensitized and perovskite solar cells and modules based on advanced nanocomposite materials. In: Printable Solar Cells. Wiley Online Library, pp 363–381
98. Wu J, Gao M, Chai Y et al (2021) Towards a bright future: the versatile applications of organic solar cells. Mater Reports Energy 1:100062
99. Fukuda K, Yu K, Someya T (2020) The future of flexible organic solar cells. Adv Energy Mater 10:2000765

100. Li Y, Xu G, Cui C, Li Y (2018) Flexible and semitransparent organic solar cells. Adv Energy Mater 8:1701791
101. Cheng P, Zhan X (2016) Stability of organic solar cells: challenges and strategies. Chem Soc Rev 45:2544–2582
102. Luo D, Jang W, Babu DD et al (2022) Recent progress in organic solar cells based on non-fullerene acceptors: materials to devices. J Mater Chem A 10:3255–3295
103. Li C, Zhou J, Song J et al (2021) Non-fullerene acceptors with branched side chains and improved molecular packing to exceed 18% efficiency in organic solar cells. Nat Energy 6:605–613
104. Chen S, Feng L, Jia T et al (2021) High-performance polymer solar cells with efficiency over 18% enabled by asymmetric side chain engineering of non-fullerene acceptors. Sci China Chem 64:1192–1199
105. Kamel MSA, Al-jumaili A, Oelgemöller M, Jacob MV (2022) Inorganic nanoparticles to overcome efficiency inhibitors of organic photovoltaics: an in-depth review. Renew Sustain Energy Rev 166:112661
106. Doumon NY, Yang L, Rosei F (2022) Ternary organic solar cells: a review of the role of the third element. Nano Energy 94:106915
107. Pal K, Singh P, Bhaduri A, Thapa KB (2019) Current challenges and future prospects for a highly efficient (>20%) kesterite CZTS solar cell: a review. Sol Energy Mater Sol Cells 196:138–156
108. Arturo M-A (2013) Thin film solar cells using earth-abudant materials. In: Solar cells: research and application perspectives, first. InTech, Rijeka
109. Khalate SA, Kate RS, Deokate RJ (2018) A review on energy economics and the recent research and development in energy and the Cu_2ZnSnS_4 (CZTS) solar cells: a focus towards efficiency. Sol Energy 169:616–633
110. Larramona G, Choné C, Meissner D et al (2020) Stability, reliability, upscaling and possible technological applications of kesterite solar cells. J Phys Energy 2:024009
111. Nugroho HS, Refantero G, Septiani NLW et al (2022) A progress review on the modification of CZTS(e)-based thin-film solar cells. J Ind Eng Chem 105:83–110
112. Crovetto A, Hansen O (2017) What is the band alignment of $Cu_2ZnSn(S, Se)_4$ solar cells? Sol Energy Mater Sol Cells 169:177–194
113. He M, Yan C, Li J et al (2021) Kesterite solar cells: insights into current strategies and challenges. Adv Sci 8:2004313
114. Alivisatos AP (1996) Semiconductor clusters, nanocrystals, and quantum dots. Science (80-) 271:933–937
115. Sogabe T, Shen Q, Yamaguchi K (2016) Recent progress on quantum dot solar cells: a review. J Photonics Energy 6:040901
116. Mora-Seró I (2020) Current challenges in the development of quantum dot sensitized solar cells. Adv Energy Mater 10:2001774
117. (2023) Canon develops perovskite quantum-dot inks for use in next-generation displays, with improved durability and potential for application in high-image-quality displays. https://global.canon/en/news/2023/20230529.html. Accessed 12 Jul 2023
118. Kongkanand A, Tvrdy K, Takechi K et al (2008) Quantum dot solar cells. Tuning photoresponse through size and shape control of $CdSe-TiO_2$ architecture. J Am Chem Soc 130:4007–4015
119. Kramer IJ, Sargent EH (2014) The architecture of colloidal quantum dot solar cells: materials to devices. Chem Rev 114:863–882
120. Carey GH, Abdelhady AL, Ning Z et al (2015) Colloidal quantum dot solar cells. Chem Rev 115:12732–12763
121. Singh S, Khan ZH, Khan MB et al (2022) Quantum dots-sensitized solar cells: a review on strategic developments. Bull Mater Sci 45:1–13
122. Peighambardoust NS, Sadeghi E, Aydemir U (2022) Lead halide perovskite quantum dots for photovoltaics and photocatalysis: a review. ACS Appl Nano Mater
123. Liu L, Najar A, Wang K et al (2022) Perovskite quantum dots in solar cells. Adv Sci 9:2104577

124. Pan Z, Rao H, Mora-Seró I et al (2018) Quantum dot-sensitized solar cells. Chem Soc Rev 47:7659–7702
125. Rasal AS, Yadav S, Kashale AA et al (2022) Stability of quantum dot-sensitized solar cells: a review and prospects. Nano Energy 94:106854
126. Wikipedia Perovskite solar cell. https://en.wikipedia.org/wiki/Perovskite_solar_cell. Accessed 23 Oct 2022
127. Perovskite-info Perovskite solar panels developers. https://www.perovskite-info.com/companies-list/perovskite-solar-panels-developers. Accessed 23 Oct 2022
128. Saikia D, Betal A, Bera J, Sahu S (2022) Progress and challenges of halide perovskite-based solar cell- a brief review. Mater Sci Semicond Process 150:106953
129. Bailie CD, McGehee MD (2015) High-efficiency tandem perovskite solar cells. MRS Bull 40:681–685
130. Qiu L, Ono LK, Qi Y (2018) Advances and challenges to the commercialization of organic–inorganic halide perovskite solar cell technology. Mater Today Energy 7:169–189
131. Wang P, Wu Y, Cai B et al (2019) Solution-processable perovskite solar cells toward commercialization: progress and challenges. Adv Funct Mater 29:1807661
132. Green MA, Ho-Baillie A, Snaith HJ (2014) The emergence of perovskite solar cells. Nat Photonics 8:506–514
133. Saule technologies. https://sauletech.com/wp-content/uploads/2019/08/saule-a4-module-1.jpg. Accessed 10 Jul 2023
134. Kim JY, Lee J-W, Jung HS et al (2020) High-efficiency perovskite solar cells. Chem Rev 120:7867–7918
135. Liu R, Xu K (2020) Solvent engineering for perovskite solar cells: a review. Micro Nano Lett 15:349–353
136. Taylor AD, Sun Q, Goetz KP et al (2021) A general approach to high-efficiency perovskite solar cells by any antisolvent. Nat Commun 12:1–11
137. Ghosh S, Mishra S, Singh T (2020) Antisolvents in perovskite solar cells: importance, issues, and alternatives. Adv Mater Interfaces 7:2000950
138. Ibrahim IAM, Chung C-Y (2022) Lead-free double perovskites: how divalent cations tune the electronic structure for photovoltaic applications. J Mater Chem C 10:12276–12285
139. Yang Z, Babu BH, Wu S et al (2020) Review on practical interface engineering of perovskite solar cells: from efficiency to stability. Sol RRL 4:1900257
140. Li H, Chen C, Hu H et al (2022) Strategies for high-performance perovskite solar cells from materials, film engineering to carrier dynamics and photon management. InfoMat 4:e12322
141. Li N, Niu X, Chen Q, Zhou H (2020) Towards commercialization: the operational stability of perovskite solar cells. Chem Soc Rev 49:8235–8286
142. Chalkias DA, Karavioti A, Papanicolaou GC, Stathatos E (2022) Stability assessment of carbon-based hole-transport-layer-free perovskite solar cells under accelerated ageing: a combined experimental and predictive modelling analysis. Electrochim Acta 427:140905
143. Ma S, Yuan G, Zhang Y et al (2022) Development of encapsulation strategies towards the commercialization of perovskite solar cells. Energy Environ Sci 15:13–55
144. Baker JA, Mouhamad Y, Hooper KEA et al (2017) From spin coating to roll-to-roll: investigating the challenge of upscaling lead halide perovskite solar cells. IET Renew Power Gener 11:546–549
145. Karavioti A, Chalkias DA, Katsagounos G et al (2021) Toward a scalable fabrication of perovskite solar cells under fully ambient air atmosphere: from spin-coating to inkjet-printing of perovskite absorbent layer. Electronics 10:1904
146. Fong PWK, Li G (2021) The challenge of ambient air–processed organometallic halide perovskite: technology transfer from spin coating to meniscus blade coating of perovskite thin films. Front Mater 8:635224
147. Mohamad Noh MF, Arzaee NA, Nawas Mumthas IN et al (2020) High-humidity processed perovskite solar cells. J Mater Chem A 8:10481–10518
148. Li D, Zhang D, Lim K-S et al (2021) A review on scaling up perovskite solar cells. Adv Funct Mater 31:2008621

149. Chalkias DA, Mourtzikou A, Katsagounos G et al (2022) Suppression of coffee-ring effect in air-processed inkjet-printed perovskite layer toward the fabrication of efficient large-sized all-printed photovoltaics: a perovskite precursor ink concentration regulation strategy. Sol RRL 6:2200196
150. Zhang L, Chen S, Wang X et al (2021) Ambient inkjet-printed high-efficiency perovskite solar cells: manipulating the spreading and crystallization behaviors of picoliter perovskite droplets. Sol RRL 5:2100106
151. Tai Q, You P, Sang H et al (2016) Efficient and stable perovskite solar cells prepared in ambient air irrespective of the humidity. Nat Commun 7:11105
152. Chalkias DA, Karavioti A, Kalarakis AN, Stathatos E (2021) Unveiling the importance of dripping temperature control of hybrid organic-inorganic perovskite precursor solution for the fabrication of fully ambient air-processed perovskite solar cells. Sol Energy 224:1017–1027
153. Pescetelli S, Agresti A, Viskadouros G et al (2022) Integration of two-dimensional materials-based perovskite solar panels into a stand-alone solar farm. Nat Energy 7:597–607
154. Kim H-S, An Y-J, Il KJ et al (2022) Sustainable green process for environmentally viable perovskite solar cells. ACS Energy Lett 7:1154–1177
155. Chalkias DA, Mourtzikou A, Katsagounos G, et al (2023) Development of greener and stable inkjet-printable perovskite precursor inks for all-printed annealing-free perovskite solar mini-modules manufacturing. Small Methods 2300664
156. Fang Z, Zeng Q, Zuo C, et al (2021) Perovskite-based tandem solar cells. Sci Bull 66:621–636
157. Li L, Wang Y, Wang X et al (2022) Flexible all-perovskite tandem solar cells approaching 25% efficiency with molecule-bridged hole-selective contact. Nat Energy 7:708–717
158. Chen X, Jia Z, Chen Z et al (2020) Efficient and reproducible monolithic perovskite/organic tandem solar cells with low-loss interconnecting layers. Joule 4:1594–1606
159. Ludin NA, Mustafa NI, Hanafiah MM et al (2018) Prospects of life cycle assessment of renewable energy from solar photovoltaic technologies: a review. Renew Sustain Energy Rev 96:11–28
160. Frischknecht R, Heath G, Raugei M, et al (2016) Methodology guidelines on life cycle assessment of photovoltaic electricity, 3rd ed. third. U.S. Department of Energy Office of Scientific and Technical Information
161. Rashedi A, Khanam T (2020) Life cycle assessment of most widely adopted solar photovoltaic energy technologies by mid-point and end-point indicators of ReCiPe method. Environ Sci Pollut Res 27:29075–29090
162. Lunardi MM, Dias PR, Deng R, Corkish R (2021) Life cycle environmental assessment of different solar photovoltaic technologies. In: Photovoltaic sustainability and management. AIP Publishing, New York
163. Hou G, Sun H, Jiang Z et al (2016) Life cycle assessment of grid-connected photovoltaic power generation from crystalline silicon solar modules in China. Appl Energy 164:882–890
164. Hong J, Chen W, Qi C et al (2016) Life cycle assessment of multicrystalline silicon photovoltaic cell production in China. Sol Energy 133:283–293
165. Akinyele DO, Rayudu RK, Nair NKC (2017) Life cycle impact assessment of photovoltaic power generation from crystalline silicon-based solar modules in Nigeria. Renew Energy 101:537–549
166. Santoyo-Castelazo E, Solano-Olivares K, Martínez E et al (2021) Life cycle assessment for a grid-connected multi-crystalline silicon photovoltaic system of 3 kWp: a case study for Mexico. J Clean Prod 316:128314
167. Luo W, Khoo YS, Kumar A et al (2018) A comparative life-cycle assessment of photovoltaic electricity generation in Singapore by multicrystalline silicon technologies. Sol Energy Mater Sol Cells 174:157–162
168. M. Lunardi M, Alvarez-Gaitan JP, Chang NL, Corkish R (2018) Life cycle assessment on PERC solar modules. Sol Energy Mater Sol Cells 187:154–159
169. Méndez L, Forniés E, Garrain D et al (2021) Upgraded metallurgical grade silicon and polysilicon for solar electricity production: a comparative life cycle assessment. Sci Total Environ 789:147969

170. Müller A, Friedrich L, Reichel C et al (2021) A comparative life cycle assessment of silicon PV modules: impact of module design, manufacturing location and inventory. Sol Energy Mater Sol Cells 230:111277
171. Lim MSW, He D, Tiong JSM et al (2022) Experimental, economic and life cycle assessments of recycling end-of-life monocrystalline silicon photovoltaic modules. J Clean Prod 340:130796
172. Klugmann-Radziemska E, Kuczyńska-Łażewska A (2020) The use of recycled semiconductor material in crystalline silicon photovoltaic modules production—a life cycle assessment of environmental impacts. Sol Energy Mater Sol Cells 205:110259
173. Kittner N, Gheewala SH, Kamens RM (2013) An environmental life cycle comparison of single-crystalline and amorphous-silicon thin-film photovoltaic systems in Thailand. Energy Sustain Dev 17:605–614
174. Mohr NJ, Meijer A, Huijbregts MAJ, Reijnders L (2013) Environmental life cycle assessment of roof-integrated flexible amorphous silicon/nanocrystalline silicon solar cell laminate. Prog Photovoltaics Res Appl 21:802–815
175. Peng J, Lu L, Yang H (2013) Review on life cycle assessment of energy payback and greenhouse gas emission of solar photovoltaic systems. Renew Sustain Energy Rev 19:255–274
176. Kim H, Cha K, Fthenakis VM et al (2014) Life cycle assessment of cadmium telluride photovoltaic (CdTe PV) systems. Sol Energy 103:78–88
177. Mohr NJ, Schermer JJ, Huijbregts MAJ et al (2007) Life cycle assessment of thin-film GaAs and GaInP/GaAs solar modules. Prog Photovoltaics Res Appl 15:163–179
178. Smith BL (2018) Development and life cycle assessment of advanced-concept III–V multijunction photovoltaics
179. Parisi ML, Maranghi S, Basosi R (2014) The evolution of the dye sensitized solar cells from Grätzel prototype to up-scaled solar applications: a life cycle assessment approach. Renew Sustain Energy Rev 39:124–138
180. Anctil A, Babbitt C, Landi B, Raffaelle RP (2010) Life-cycle assessment of organic solar cell technologies. Conf Rec IEEE Photovolt Spec Conf 742–747
181. Tsang MP, Sonnemann GW, Bassani DM (2016) Life-cycle assessment of cradle-to-grave opportunities and environmental impacts of organic photovoltaic solar panels compared to conventional technologies. Sol Energy Mater Sol Cells 156:37–48
182. Mokhtarimehr M, Forbes I, Pearsall N (2018) Environmental assessment of vacuum and non-vacuum techniques for the fabrication of Cu_2ZnSnS_4 thin film photovoltaic cells. Jpn J Appl Phys 57:08RC14
183. Resalati S, Okoroafor T, Maaloufa A et al (2022) Life cycle assessment of different chalcogenide thin-film solar cells. Appl Energy 313:118888
184. Celik I, Song Z, Cimaroli AJ et al (2016) Life cycle assessment (LCA) of perovskite PV cells projected from lab to fab. Sol Energy Mater Sol Cells 156:157–169
185. Correa Guerrero NB, Herrera Martínez WO, Civit B, Perez MD (2021) Energy performance of perovskite solar cell fabrication in Argentina. A life cycle assessment approach. Sol Energy 230:645–653
186. Okoroafor T, Maalouf A, Oez S et al (2022) Life cycle assessment of inkjet printed perovskite solar cells. J Clean Prod 373:133665
187. Gong J, Darling SB, You F (2015) Perovskite photovoltaics: life-cycle assessment of energy and environmental impacts. Energy Environ Sci 8:1953–1968
188. Tian X, Stranks SD, You F (2021) Life cycle assessment of recycling strategies for perovskite photovoltaic modules. Nat Sustain 4:821–829

Chapter 5
Designing the Future of Agrivoltaics

This chapter is a consolidated map of the future of agrivoltaics strategy, presenting the most significant emerging agrivoltaic concepts, from their design level to practical realizations, that appeared in the last few years. Emphasis is given to novel agrivoltaic concepts employing both the mature crystalline silicon photovoltaics and emerging solar cell technologies, with the latter to promise a revolution in land-sharing approach, aimed at more efficient and sustainable energy production and crop growth, simultaneously. Finally, a short discussion providing a comparison of the characteristics of the different photovoltaic technologies tested until now in agricultural practices, focused on important criteria for this niche application, is presented.

5.1 Introduction

Technological advancements have long been viewed as the driving force behind the development of humankind and a valuable tool in meeting its ever-increasing needs. Humans have always been innovative in managing their demands, with the progress pace being determined by the number and criticality of the challenges faced in each time era. At the United Nations Scientific Conference for the Conservation and Utilization of Resources in 1949, Trygve Lie, the United Nations Secretary-General, launched the meeting with an optimistic appraisal [1]:

> Modern science has developed a wide range of new techniques for wresting more wealth from the Earth's resources. Their effective application is basic to achieving a world of abundance - a world where the specters of hunger and wastelands will be things of the past.

Today, more than ever, the Water-Energy-Food-Ecosystems (WEFE) nexus faces the grandest challenges that undermine the prosperity and sustainable development of life on Earth. Emerging technologies are expected and should be key players in this run, and so be this in the energy and agricultural sectors. Agrivoltaic practices

D. A. Chalkias and E. Stathatos, *The Emergence of Agrivoltaics*,
Green Energy and Technology, https://doi.org/10.1007/978-3-031-48861-0_5

have increasingly been seen as one of the most effective strategies and "single-step" solution to tackle WEFE nexus challenges, alleviating the land-use competition between energy and food production, as well as increasing land efficiency. However, important barriers are still faced in their adoption, with many of them concerning their technological maturity, highly hindering their widespread application finally. This has awakened researchers to further develop or even relaunch the concept of agrivoltaics, demonstrating novel and creative ideas, and more effective foundations in the field.

5.2 Novel Agrivoltaic Concepts Employing the Mature Crystalline Silicon Solar Cells

Although crystalline silicon (c-Si) photovoltaics (PVs) are the most mature devices for light-to-electricity conversion, providing the lowest levelized cost of electricity (LCOE), there are a number of limitations for their widespread adoption in agricultural practices, mainly stemming from specific attributes. These include their rigidity, heavy-weight and relatively low performance in diffuse light conditions, while the opacity that characterizes these systems is considered of the most important barriers for this specific application, highly limiting the growth of plants in the case of cultivation underneath them. Nevertheless, the employment of conventional c-Si PVs is still considered the dominant choice in the agrivoltaics market at the moment. This is mainly realized after light management using two major trends [2]. The first one includes the application of conventional PV modules, like those found in rooftops and solar power plants around the world, spaced apart from each other so as to allow incoming solar light to reach the crops on the ground. The other case and most effective one regards the usage of semi-transparent PV modules designed and manufactured in such a way that their surface is only partially populated with solar cells, allowing transparent spaces within each module for light to be able to reach the plants under them (see Fig. 5.1).

Power capacity and transparency at the module level can be manipulated depending on the specific practice (e.g., type of the crop, geographical location of the cultivation, etc.). To that aim, geometric modeling is nowadays used as a quite valuable tool, helping to determine the interrelation of the involved parameters under a multi-scale (module-string-array) parametrization, which will finally determine the agrivoltaic topology [4]. More specifically, starting from the "micro" scale, solar cells dimensions, number and spacing, are initially all parametrically defined. Front and rear covers, as well as the frame of single modules, are also considered. At the "meso" scale, inter-column-spacing (ICS) and inter-row-spacing (IRS) are set. In the end, at the "macro" scale, the row spacing (RS) and column spacing (CS) are configured to convert the strings of modules into arrays, finally reaching a suitable agrivoltaic design (see Fig. 5.2).

Fig. 5.1 An example of an agrivoltaic practice based on crystalline silicon solar cell technology with modulated transparency at the module level [3]

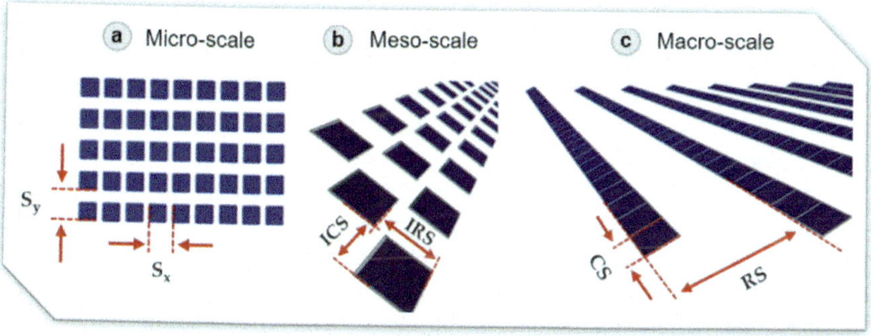

Fig. 5.2 Multi-scale parametrization of agrivoltaic topologies: **a** single module, **b** string and **c** array [4]. (Reproduced from Ref. [4], https://ars.els-cdn.com/content/image/1-s2.0-S0306261921016986-gr3_lrg.jpg, under the terms of the CC BY 4.0 license)

Following the continuous development of c-Si PV technology, specialized solutions with increased benefits in the simultaneous light-to-electricity conversion and plants cultivation can now be adopted for each agrivoltaic practice. Today, the modern bifacial PVs are considered ideal for agrivoltaic practices that aim for high transparency but, at the same time, increased electric power production. Researchers in the field have found that bifacial PV arrays are prominently advantageous in stilt-mounted projects, where modules are elevated considerably off the ground at over 2 m, thus allowing plenty of light to be absorbed by the rear PV side as well [5]. For light-demanding crops, where the amount of incident irradiance on the ground should be abundant, highly transparent bifacial PVs set an ideal synergy between cultivation and electricity production, offering an increased power capacity for similar-sized agrivoltaic systems compared to the mono-facial counterparts, which can even reach the level of 40% [4].

The great variety of possible requirements in terms of light transparency and power production means that a large number of different semi-transparent PV module configurations are needed in order to be able to offer an optimal agrivoltaic solution for different scenarios. However, in this case, it is the production lines of the PV modules that become a limiting factor. Most of the production lines are designed for manufacturing modules that are fully populated with solar cells, with a lack of flexibility for shifting the product configurations, as required in agrivoltaic practices. In this case, the manufacturing of agrivoltaic modules introduces inefficiencies in the production capacities and costs, curtailing the feasibility of a wide deployment of agrivoltaic products. In this direction, today, more and more industrial companies are working on the insight of developing manufacturing lines with production flexibility for various module configurations, studying the attributes of their products for the best-case scenario results thoroughly.

Although the concept of partial PV roof covering is effective during sunny days, shading becomes excessive for cultivation during cloudy days and low-irradiance periods. Adjusting the shading level by rotating the roof-integrated PVs has been proposed by many researchers in the field. This strategy has already demonstrated quite beneficial results for agricultural activities in greenhouses since, in this way, the internal microclimatic parameters can be manipulated by changing the roof shading levels. Marucci et al. developed a PV greenhouse based on this strategy using PV modules and mirrors (see Fig. 5.3). In one of their studies, they tested the system performance by regulating the shading percentage inside the greenhouse to be 30% in winter and 80% in summer. The presence of mirrors allowed high electricity production, even with non-optimal PV system angles. They demonstrated that by lowering solar radiation available inside the structure by shading under the PV modules, increased electricity was produced, while this also led to a decrease of the air temperature inside the greenhouse, working as a passive cooling system [6]. This is considered very important, especially for geographical areas characterized by high solar irradiation.

In the search for attaining more efficient PV modules without sacrificing their initial transparency, additional novel concepts have already been proposed, with one of them being the technology of luminescent solar concentrators (LSCs) [8]. LSCs is a versatile and powerful photonic technology owing to their ability to efficiently concentrate the incident light (even diffuse light) to light of a specific wavelength. More specifically, the light spectrum regions that are absorbed by the LSC compounds are re-emitted by down-conversion (conversion of one photon of a higher energy into a pair of photons of lower energy) or up-conversion (multiple low-energy photons

Fig. 5.3 Prototype photovoltaic greenhouse with variable roof shading for optimization of agricultural and energy production [7]. (Reproduced from Ref. [7], https://www.mdpi.com/1996-1073/12/13/2589, under the terms of the CC BY 4.0 license)

are converted in higher-energy photons) processes to achieve the desired light wavelengths. The LSC compounds are specially developed compounds, organic, rare-earth complexes or quantum dots, commonly structured in the form of a coating on a transparent substrate. For agrivoltaics applications, an LSC can be applied as a coating on the front side substrate of a PV module to capture non-critical or even harmful portion of light for both PVs and plants and re-emit it to useful light wavelength for both solar cells operation and photosynthetic action of the plants. In this way, the performance of solar cells and the light transmittance of a PV module in critical wavelengths of light for the plants growth are increased, while both the PV system and cultivation are protected from the harmful portion of sunlight (i.e., ultraviolet light). The technology of LSCs has already been applied in prototype PV greenhouses, while the first commercially available products are also entering the market.

The concept of LSC based on organic dyes as relevant luminescent species was introduced in 1976. Weber et al. [9] first demonstrated that dye-doped glasses had the potential to function as greenhouse luminescent sensors with an estimated efficiency value of 10%. After then, many different organic dyes synthesized from coumarins, rhodamines and perylene derivatives have drawn significant attention as promising luminescent materials for LSC devices [10]. These dye categories show an absorption spectrum between 350 to 580 nm, whereas their emission spectrum ranges from 400 to 600 nm. Although a significant overlap of the absorption and emissions values is indicated, the quantum yield efficiency of these materials exceeds 85%, matching the light spectrum that is highly useful for both solar cells operation and the photosynthetic action of the plants. Today, LSC technology is entering the agrivoltaic market. As an example, Soliculture is a company dedicated to the development of PV greenhouses employing dye-based LSC technology (see Fig. 5.4). Soliculture has already installed both commercial-scale and research LSC energy systems over various locations to test plant responses in a range of climates. The company develops

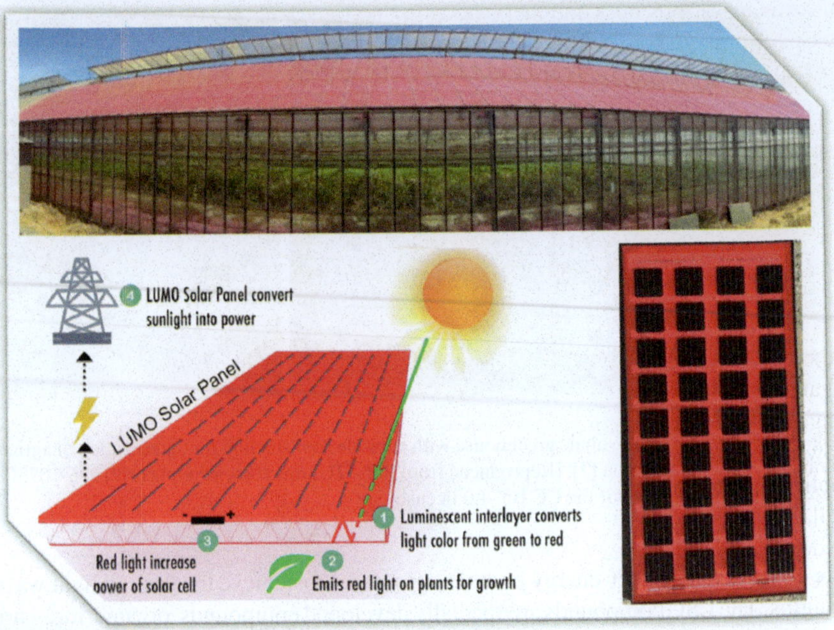

Fig. 5.4 Photovoltaic greenhouse employing the technology of luminescent solar concentrator and a schematic representation of the technology [11]

semitransparent solar panels containing a low density of silicon PV strips arranged periodically on a panel of glass, allowing light to transmit between the strips. Solar modules of different maximum power and transparency are produced, depending on the agrivoltaic project, with the power conversion efficiencies (PCEs) to range between 5 and 12%, achieved with high-efficiency bifacial passivated emitter and rear contact solar cells. On the front-side of the modules, a thin layer of luminescent material (light-altering dye) is adhered to the backside of the glass substrate, enhancing light quality by converting green light to red light. This optimized light spectrum enhances power production and facilitates plant growth [11].

Although many LSC developers have primarily focused up-to-now on organic dye-based systems, specific challenges are faced by the use of these materials, some of them concerning re-absorption, photo-stability and low Stokes-shift. Consequently, the scientific community is currently focused on alternative materials for the development of LSCs. Rare-earth complexes possess some excellent characteristics compared to organic dyes, such as high photo-stability, high absorption coefficients, fewer re-absorption losses and significant Stokes-shift (around 300 nm) [10]. By selecting an appropriate luminescent rare-earth complex, which absorbs radiation in the non- photosynthetic active radiation (PAR) spectral region, efficient crop growth can be achieved. For example, LSCs based on rare-earth complexes mixed with various host polymers are transparent, which is essential in the greenhouse for daylight and low electricity consumption. Lanthanides based on Eu^{3+} and Tb^{3+} ions are commonly employed for this aim [12]. However, for these materials, further improvements in their quantum yield should be achieved to attain the desired results, which prior require a deeper understanding of their molecular structures. The development of innovative luminescent rare-earth, with increased performance in the visible spectrum of light, would be highly advantageous for efficient crop growth as well as electrical power in agrivoltaic technologies.

Except for some dyes and rare earth complexes that selectively absorb either visible or ultraviolet light, respectively, and re-emit it to specific wavelengths, very small-sized semiconductors, such as quantum dots (QDs), can also participate in the mechanism of LSCs [13]. QDs can offer additional advantages over the typical organic chromophores used in LSC technology, such as a wider spectral absorption range, straightforward tunability of the emission and potential for color neutrality. However, among the main drawbacks of usage of this class of materials for this application are their high manufacturing cost, toxicity issues for some of them, as well as that they do not exhibit bright near-infrared emission and large Stokes shift. To deal with the aforementioned drawbacks, scientists are nowadays working on the development of new materials that have the potential to achieve the desired attributes. An example is the development of $CuInS_2$ QDs. These materials can be developed under low-cost, simple heat-up methods, do not present toxicity issues, have a large Stokes shift and can be fabricated with almost 100% photoluminescence quantum yield in the near-infrared spectral range [14].

Another strategy that can provide a positive impact on light management for agrivoltaic practices and that moves beyond the state-of-the-art in the field, is the utilization of anti-reflective (AR) coatings for increased light transmittance [15]. AR coatings are widely used to increase the transmission of light and reduce glare in a diverse range of glass applications, including architectural and automotive glass, electronic displays and energy devices. Today, the scientific community aims at an even wider application range of this technology (including cost-sensitive industries) by considering economic aspects as top priorities. High-cost manufacturing techniques (e.g., vacuum deposition) are now investigated to be replaced by lower-cost and scalable ones (e.g., printing deposition). On the other hand, novel nanomaterials are designed to offer the desired attributes at a lower cost. Regarding the agricultural sector, there has been an increasing number of investigations in the last few years focusing on the development of low-cost AR coatings for greenhouses integration. Dey et al. [16] developed a nano-porous sol–gel derived hydrophobic glass coating for increased light transmittance through the greenhouse. Their results showed an increased light transmittance on the order of 7% on the whole visible spectrum of light, which can benefit both solar cells operation and plants growth. The coating also demonstrates a very high scratch resistance, which is very important for real applications. Wu et al. [17] also proposed an anti-reflective film to improve the photosynthetic spectral transmittance in a PV greenhouse (see Fig. 5.5). In their work, they found that the whole-day average transmissivities of the spectral splitting covering (SSC) coated with the anti-reflective film at 450 and 660 nm were

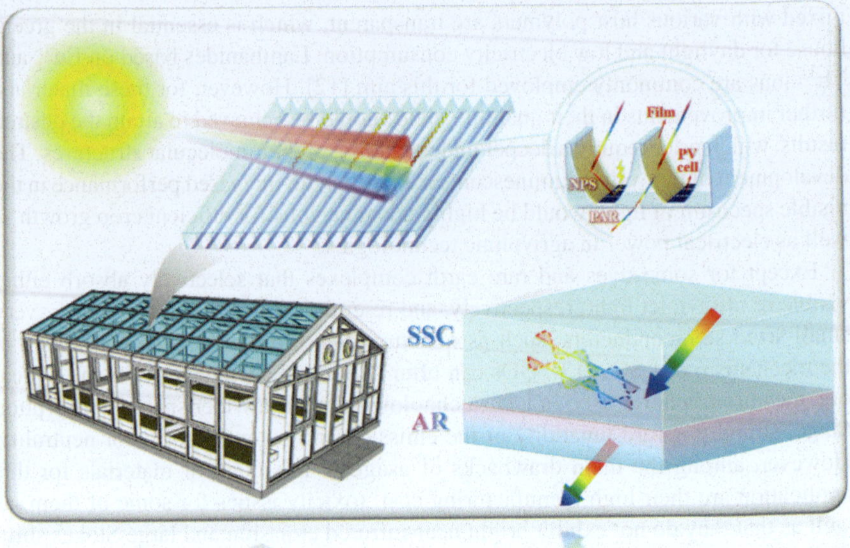

Fig. 5.5 The working principle of spectral splitting covering coated with anti-reflective coating [17]

57.6% and 61.5%, respectively. The whole-day average crop growth factor with the usage of the anti-reflective film was found at 56.86%, higher than its reference value (24.8%), which represents an important increase in the active light flux entering the greenhouse for the plants growth.

Finally, novel semi-transparent PV products based on c-Si technology have appeared for the agricultural sector in the last few years, revealing alternative agrivoltaic concepts for efficient dual-land use. Amongst them, of particular interest are the semi-transparent PV modules based on c-Si micro solar cells (see Fig. 5.6). This type of modules has already been tested for greenhouses roof application by Yano et al. [18, 19]. Under this approach, the benefit is that the solar radiation is distributed homogeneously in the greenhouse area because the dimensions of the cells are not enough to shade the sun reaching the plants. This feature is distinguished from the translucency obtained by the PV modules partially populated with planar c-Si solar cells, where the sun is at an important level obscured. Additionally, due to the unique geometry of these solar cells (spherical type), sun rays coming from any spatial direction can be utilized, even those coming from the ground reflection, thus producing a constant amount of energy at a wide range of sunlight incidence angles. This attribute makes the semi-transparent spherical micro-cell technology suitable for application even at the sidewalls of greenhouses.

Fig. 5.6 Semi-transparent photovoltaic module for greenhouses application based on spherical crystalline silicon micro-cells of 1.8 mm diameter [18]. (Reproduced from Ref. [18], https://ars.els-cdn.com/content/image/1-s2.0-S1537511014000580-fx1.jpg, under the terms of the CC BY 3.0 license)

5.3 Sunsharing Revolution in Agriculture Through the Application of Commercialized Thin-Film, III–V Multi-junction or Emerging Solar Cell Technologies

Recognizing the growing interest in the development of agrivoltaic systems as a valuable strategy in the direction of sustainable management of the WEFE nexus, next-generation solar cell technologies are increasingly tested in electric power production combined with agricultural practices. Leaning on the very high-efficiency of III–V multi-junction solar cells, as well as the unique attributes that characterize the second-generation and emerging solar cell technologies, the aim of their employment in agriculture is to overcome the barriers that appeared by the usage of the conventional planar c-Si PVs usually applied in the agrivoltaic systems up until today and proceed to a sunsharing revolution with new milestone achievements in the simultaneous energy production and plants cultivation dual-land use.

Considering the very high efficiency of III–V multi-junction solar cells, a new insight for developing high performance and transparent agrivoltaic systems has emerged. More specifically, using this PV technology, the required PV area for electricity production is highly reduced, offering transparency in the space in between, in order for sunlight to reach the underneath crops to photo-synthesize. Concentrators can help even more in this direction. Optical reflectors or refractors can be used to concentrate the sunlight on the solar cells. The main advantage of those systems is that the concentration of light reduces further the required PV area, allowing the use of expensive, highly efficient III–V multi-junction solar cells. However, an important issue that usually arises for concentrator PV systems efficient operation is that they usually require a tracking system to collect sunlight efficiently since they mainly rely on direct solar light conversion to electricity. For this reason, this solution is more economically attractive for regions with direct normal irradiation of more than 2000 kWh/m^{2}a [20]. Up-to-now, different highly transparent concentrator PV modules have been designed and tested for efficient dual-land-use applications. Sato et al. [21] recently studied two types of highly transparent concentrator PV modules employing III–V triple-junction solar cells that separately use direct and diffuse sunlight (see Fig. 5.7). Type A featured an open-structure design with a typical-scale lens and solar cells, as well as a completely direct-diffuse-separated design (direct sunlight was concentrated on the solar cells for power production, while diffuse sunlight was transmitted through the module to the land beneath). Type B featured a closed-structure design with microscale lens and solar cells, as well as a flexible design in that the amount of direct sunlight transmitted can be varied by changing the ratio of the lens aperture area to the module aperture area. In this study, they demonstrated transmitted-irradiance-to-global-normal-irradiance ratios of 15.3–63.7% for type A and 38.0–63.8% for type B, respectively, for a diffuse-to-global ratio of 0.23–1.0. Their findings showed that irradiance of at least 160 W/m^2 and illuminance of at least 20 kilolux could be attained beneath the modules even if both concentrator PV modules were densely placed with no gaps between them (i.e., the ground coverage ratio is 1). These performance characteristics are significantly better than those of conventional partially transparent flat PV modules with 17%-efficiency opaque silicon solar cells. Regarding the practical application of solar concentrators in agriculture, the Insolight company established in Switzerland has already started the commercialization of a micro-concentrator module based on high-efficiency III–V multi-junction solar cells that incorporates planar micro-tracking [22].

On the other hand, translucent solar cell devices with engineered optical characteristics that their application to agricultural structures can provide precise control of the lighting conditions have today seized paramount attention in the field. The development of translucent PVs can be based on both second-generation solar cell technologies and third-generation nano-photovoltaics, with the latter group being the predominant choice for this search considering their unique and easily adjustable attributes, as well as the recent advances in their manufacturing and upscaling. Dye-sensitized solar cells (DSSCs), organic photovoltaics (OPVs) and perovskite solar cells (PSCs) are the main emerging technologies investigated for that aim. These solar systems are not only evaluated regarding the "golden triangle" requirements, i.e., their performance, stability and cost, but also using metrics that can provide a value on the influence of their application/integration as cladding materials in agricultural systems. In this respect, crop growth factor (CGF) is usually applied to evaluate the impact of PV transparency as a cladding material on the plants growth, determining the ratio of the rate of the photosynthetic action of the plants under the PVs and the rate of the photosynthetic action of the plants under the full incident sunlight [23, 24]. CGF can be derived from Eq. 5.1, where $T(\lambda)$ is the transmittance

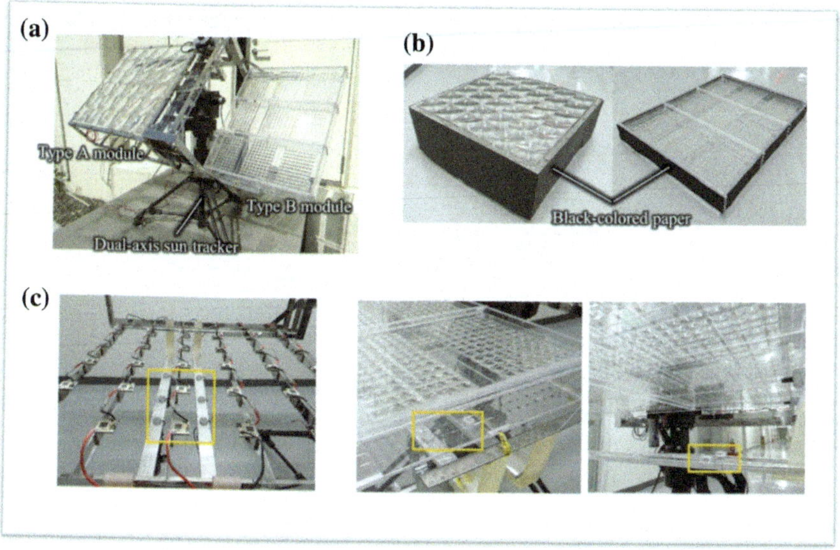

Fig. 5.7 Photograph of an outdoor experimental setup of highly transparent concentrator photovoltaic modules employing III–V triple-junction solar cells: **a** overview of experimental setup, **b** module sides covered with black shields during measurements and **c** pyranometers for sunlight transmission measurements [20]. (Reproduced from Ref. [20], https://onlinelibrary.wiley.com/doi/full/10.1002/ese3.550, under the terms of the CC BY 4.0 license)

of the semi-transparent solar cell, $b_s(\lambda)$ is the AM 1.5 solar irradiance and $a(\lambda)$ is the action spectrum of plants (usually McCREE curve is applied) [25].

$$G = \frac{\int T(\lambda) b_s(\lambda) a(\lambda) d\lambda}{\int b_s(\lambda) a(\lambda) d\lambda} \tag{5.1}$$

Regarding the second-generation PVs, semi-transparent PV glass developed using amorphous silicon (a-Si) technology has already been tested in the development of prototype PV greenhouses in the last few years. This type of solar systems can be applied to the whole surface of a greenhouse roof (PV coverage (PV_R) of 100%), as well as to the side windows of greenhouses, without highly affecting the plants growth underneath them as in the case of usage of opaque c-Si solar cells. A study carried out in the field has shown that even with a PVR of 100% (see Fig. 5.8), a slight reduction of the biomass production of the plants was shown, while the apical growth mechanism of heliophilic plants was accelerated (for horticultural crops). Nevertheless, to obtain results with a greater scope and practical application, it is necessary to extend this study to different crops and growing seasons [26].

Regarding third-generation nano-photovoltaics, DSSCs are considered the most mature in research and development, and very promising PV technology for application in the agricultural sector with great insights for milestone achievements in this field. Since each material applied in these devices is responsible for a different

Fig. 5.8 Photovoltaic greenhouse developed using amorphous silicon solar modules (30% transparency) in Spain [26]. (Reproduced from Ref. [26], https://www.mdpi.com/2073-4395/11/6/1097, under the terms of the CC BY 4.0 license)

function in the photo-electric effect, a DSSC designer can fine-tune each material characteristics for this purpose. In this way, there is great versatility in the optical characteristics of these devices, which is considered a very important attribute for the design of the next generation agrivoltaics. Until today, an important number of quite interesting investigations aimed at DSSCs application/integration in agricultural practices has already been published, with most of them being focused on closed-field agricultural practices (i.e., PV greenhouses development). Valuable PV system prototypes are demonstrated by employing novel nanostructures and dye-sensitizers that can lead to increased transparency in the PAR region. Focused on the heart of this technology, which is the dye-sensitizer, different types of organic and metal-based dye molecules have been developed to serve as the main active material. New dyes with engineered optical characteristics to provide a high transparency in the PAR region but at the same time achieve a high performance have already been designed, realized and implemented in this type of solar cells from different research groups. In this category of solar cells, PCEs of up to 8.77% have already been reported using newly synthesized organic dyes, while strategies employing binary spectral engineering modifications, such as the replacement of both conventional dyes and electrolytes with new materials with customized characteristics for agriculture, demonstrate an increased interest [24, 27, 28]. On the other hand, under the progress achieved in the last years in industrial manufacturing of this solar cell technology, several prototype dye-sensitized solar modules (DSSMs) have already been developed to be used as a large-scale cladding material for greenhouses. Noteworthily, Brite Solar company in Greece has succeeded the development of a prototype PV greenhouse of 100 m^2 size, fully covered by 2500 cm^2 semitransparent DSSMs, showing a transparency at the red region of the visible spectrum higher than 60% (see Fig. 5.9). Another interesting project has been carried out in Italy,

Fig. 5.9 Prototype photovoltaic greenhouse developed by Brite Solar company using dye-sensitized solar modules in Greece (top) and dye-sensitized solar panel applied in an aquaponic greenhouse in Italy [29, 30]. ((Bottom) is reproduced from Ref. [30], https://www.mdpi.com/1996-1073/14/19/6393, under the terms of the CC BY 4.0 license)

where 2 m^2 dye-sensitized solar panels (DSSPs) were developed for their application inside an aquaponic greenhouse. The PVs showed noteworthy efficiency-transparency combined-performance attributes, while the solar cells were also tested for a prolonged period in their aging, showing promising results.

Simultaneously, interesting experimental works are continuously published to validate DSSCs technology real application in agriculture and reveal the effect of their usage as a cladding material on plants growth. In this direction, a case study exploring the impact of the usage of this type of PVs on hydroponic cultivation of two different tomato crops grown in greenhouses is reported by Kadoglidou et al. [31]. This group compared the growth of medium-sized and cherry tomato crops in two experimental greenhouses, one conventional structured and another employing DSSCs as the covering material. The acquired data concerned the microclimate conditions inside both greenhouses (i.e., solar radiation, air temperature and humidity), average fruit weight and yield, as well as chemical analyses of harvested tissues and fruits in order to determine potential differences in the concentration of phytochemical compounds. The results showed that tomato plants grown in the DSSC-based greenhouse presented lower early and total yields, as well as lower chlorophyll content, stomatal conductance, photosynthetic rate and transpiration rate values, especially in the case of medium-sized fruits, as compared to the conventional greenhouse. On the other hand, the crops grown in the DSSC-based greenhouse showed significantly higher values of bioactive compounds, with increases in the ascorbic acid, lycopene, β-carotene and total carotenoids concentration. Finally, for both types of tomatoes, the 2,2′-azino-bis-3-ethylbenzthiazoline-6-sulphonic acid (ABTS) and 1,1-diphenyl-2-picrylhydrazyl (DPPH) tests showed circa 10% and 5% increase, respectively, in the DSSC-based greenhouse compared to the case of a conventional greenhouse.

On the other side, OPVs emerge as a very promising alternative solar cell technology for application in agricultural practices. This technology is particularly attractive for greenhouses application since it can be designed to provide high transparency in the visible spectrum of light but at the same time be effective in conversion of the rest of the light to electricity. This can be achieved by the proper selection and engineering of light-active materials (a multitude of organic materials is now available for this application), which can enable a strong light absorption in the near IR region but at the same time allow visible light to pass through unhindered. Furthermore, given that these devices are amenable to low-cost and high-throughput production methods, as well as they are usually developed on thin plastic substrates, they can provide a series of advantageous features compared to conventional c-Si solar cells, such as flexibility and light-weight with a competitive LCOE.

Today, multiple research groups and institutes are working on the development of semi-transparent OPVs for greenhouses application. Under the rapid advancements in non-fullerene small molecule electron acceptors that led to a great increase of PCE records in this technology, OPVs have attracted a great deal of interest from the scientific community again, focused now on spectral engineering strategies on these devices for niche applications [32]. In the last years, flexible semi-transparent organic solar cells have demonstrated highly improved performance attributes that give new insights in power-crop production dual-land use. PCEs on the order of 10% from devices that show an average visible light transmittance (AVT) of over 20% have already been reported in the field, while even better combined-performance attributes have very recently been achieved by OPVs developed on surface conductive glass substrates (PCE > 12% with AVT > 36%) [33, 34]. On the other hand, important steps have been made in the practical realization of OPVs-based agrivoltaics. Prototype PV greenhouses employing OPVs have already been developed in different regions around the world, such as in Ireland and the USA (see Fig. 5.10), while investigations demonstrating the effect of OPVs usage as a cladding material in greenhouses have been increasingly published in the last years. Teitel et al. [35] examined the usage of

Fig. 5.10 Prototype organic photovoltaic greenhouse developed in Ireland by ASCA (top) and in the USA by ARMOR solar films (bottom) [36, 37]. Top image credits: David Beattie, future light from distant stars 2022–23, VISUAL Carlow. Funded by the ESB brighter future arts fund, in partnership with business to arts. Photo: Ros Kavanagh 2022. (The bottom image was reproduced from Ref. [36], https://www.mdpi.com/2624-7402/4/4/62, under the terms of the CC BY 4.0 license)

flexible semi-transparent OPV modules as shading elements (covering 37% of the roof) for a greenhouse tunnel cultivating tomatoes in the summer growing seasons in 2018 and 2019. To this aim, a similar adjacent tunnel was also developed as a control greenhouse covered only by a polyethylene sheet in 2018 and by a polyethylene sheet and black shading screen (25% shading) in 2019. Regarding the first season of experiments, they found that the usage of OPVs in the greenhouse did not significantly influence plant development. Even though the solar radiation inside the OPVs-based tunnel was decreased, the cumulative number of tomatoes, their mass and average tomato mass were higher by 9, 36 and 21% respectively, than in the case of the control tunnel. The reason behind this observation was the much lower canopy temperature and slightly lower air temperature inside the OPV greenhouse. On the other hand, regarding the second season of experiments, the addition of a 25% black shading screen on the roof of the control tunnel resulted in a similar microclimate to the corresponding of OPVs-based tunnels, with similar canopy temperature. As a result, the cumulative number of tomatoes, their mass and average tomato mass were found also similar in the two tunnels. Thus, their experiments showed that OPVs can act as shading elements without any significant damage to the tomato plants.

Simultaneously, theoretical studies using simulated data have been increasingly published the recent years to estimate the worldwide potential for semi-transparent OPVs application to greenhouses. One of these studies that offers a direct comparison of the application of OPVs technology to the conventional c-Si one was carried out by Dipta et al. [38]. This study hypothesizes the case of development of two different greenhouses in Australia, cultivating inside tomato crops. The first one regarded a greenhouse employing c-Si PVs of 46.2 kW installed capacity, while a greenhouse employing OPVs of 18.48 kW installed capacity was the second case (panel area of the OPV greenhouse was larger, but OPVs demonstrated a lower performance, set at 9.4% PCE with 24.6% AVT). The main result produced in this work showed that the OPV greenhouse would attain 46% more above-ground dry weight of tomato crops than the standard PV greenhouse case. However, the former practice would not offer massive potential in the short term due to less revenue. The exact results and the potential of OPV greenhouses would also depend on energy and crop prices, except for other parameters. Another study that examined the life cycle environmental and economic impacts of integrating semitransparent OPVs into greenhouse designs in the replacement of conventional greenhouses with and without an adjacent solar PV array showed clear superiority of OPVs-based strategy (tomato was selected as a crop) [39]. The greenhouse that integrates the OPVs can achieve significant reductions in environmental impacts, particularly in regions with high solar insolation and electricity-intensive energy demands. As an example, in Arizona, global warming potential values for a conventional, adjacent c-Si PV and OPV greenhouse were found to be 3.71, 2.38 and 2.36 kg CO_2 eq/kg tomato, respectively.

Finally, PSCs have seized the attention of the scientific community in the last years, given their rapid and continuous development. Many agrivoltaic designers now consider perovskites as the future PV material to replace c-Si in agrivoltaic practices. To that aim, a significant advantage of PSCs compared to other emerging PV technologies is the ease of tuning of their optical characteristics. By adjusting the halide elements (and their ratio in mixed halide systems) for the development of the perovskite crystals, the bandgap of the perovskite material can be tuned from 1.1 to 2.3 eV. This feature can lead to properly designing PSCs and rationally optimizing their optical characteristics for greenhouses application [40]. Today, highly efficient semi-transparent PSCs have already been demonstrated in the field. Liu et al. [40] achieved the fabrication of 7.51%-efficient semi-transparent PSCs (40% transparency). These solar cells were able to absorb sunlight up to 550 nm, while offering the lower energy photons to pass through and contribute to the photosynthetic action of the underneath plants. Furthermore, another important characteristic of PSCs for greenhouses application is that these devices can achieve a quite high quantum efficiency even when manufactured on flexible polymer substrates. This feature opens the way for their application in greenhouses with curved roofs. In this direction, Liu et al. [41] achieved the development of a highly transparent flexible PSC (average transmittance of 60% in the region of 540–760 nm) that attained almost 8% PCE. Their investigation ended up noting that a semi-transparent flexible perovskite-polymer tandem solar cell could be developed, demonstrating 11% PCE and 50–60% transparency in the 540–700 nm wavelength range.

5.4 Comparison of the Different Photovoltaic Technologies for Their Implementation in Agricultural Practices

Until today, wafer-based c-Si solar cells dominate the PV market, which is also the case for the agrivoltaic market. Nowadays, c-Si PVs present a satisfactory performance in light-to-electricity conversion (up to 26% under direct sunlight illumination), outstanding stability (warranties spanning 20–30 years) and a low LCOE (estimations now demonstrate values of 3c€/kWh) [42–44]. Simultaneously, important progress has been achieved on the decrease of the environmental footprint of this technology, with the strategy of recycling end-of-life c-Si PV modules to get the last years increased attention [45]. However, it is a series of attributes that yet undermine the universal adoption of c-Si PV technology in agrivoltaic practices, where the aim is efficient clean-power production and plants cultivation simultaneously. Amongst the most important ones are the opacity, rigidity, low performance in diffuse light conditions, heavy-weight, and restrictions in size and shape, which characterize these solar cells. To deal with these challenges, next-generation PV technologies are increasingly tested nowadays as alternatives to develop novel agrivoltaic systems, offering unique attributes that would allow a sun-sharing revolution.

Until today, both scientists and process engineers in laboratories and industries have tested the possibility of applying/integrating alternative PV technologies to agriculture. High performance PVs (e.g., III–V, tandem cells) and 2nd/3rd generation solar cell technologies (e.g., a-Si, DSSCs, OPVs, PSCs) are increasingly implemented in agricultural practices, investigating different concepts from their design level to practical realizations, revealing new sustainable solutions for WEFE nexus

Table 5.1 Comparison of the different solar cell technologies using important criteria for their implementation in the agricultural sector. The "+" and "–" symbols denote a comparable better and lesser fulfillment of the respective criterion, while the "✓" and "x" symbols denote fulfillment and not of the respective criterion. The baseline is set to the characteristics of crystalline silicon solar cells ("o" symbol)

Solar cell technology	c-Si (monofacial/ bifacial)	III–V/tandem (concentrator or not)	2nd generation (a-Si, CdTe)	3rd generation (DSSCs, OPVs, PSCs)
LCOE criteria				
Efficiency	o	+	–	– to o
Stability	o	o	– to o	–
Cost	o	–	+	o to +
Performance under diffuse light conditions	o	o	o	+
Design criteria				
Transparency	x	x	✓	✓
Variability in size and shape	x	✓	✓	✓
Mechanical flexibility	x	x	✓	✓
Non-hazardous materials	o	o	– to o	o to +
Future potential				
Market penetration	o	–	–	–
Research activity	o	– to o	– to o	+

management. For readers' benefit, Table 5.1 presents a comparison of the most important characteristics of solar cells for their application in the agricultural sector [46].

As it is perceived, III–V/tandem PVs and next-generation solar cells provide attributes that outweigh in some criteria the characteristics of c-Si solar cells for their implementation in the agricultural sector. More specifically, III–V/tandem PVs provide a higher light-to-electricity conversion efficiency with almost the same operational stability compared to c-Si solar cells. Thus, the former type of solar cells can be employed to fabricate modules composed of low-populated solar cells, providing almost the same power production while offering at the same time increased transparency that is important for the photosynthetic action of the plants. However, as in the case of c-Si solar cells, the non-homogeneous illumination beneath the module, the low performance in diffuse light conditions, the rigidity and toxicity still remain important issues, restricting their future potential for higher market penetration. On the other side, 2nd and 3rd generation PVs offer important design benefits for their implementation in agricultural practices, with the most critical being the transparency that can be attained in the active area of the PVs. Considering this attribute, these devices can be applied as a cladding material with high PV_R in agricultural practices, providing a homogeneous illumination beneath the module. Besides, 3rd generation PVs are highly capable of spectral engineering using simple roots in order to achieve manipulated optical characteristics that aim to minimize their effect on crop growth underneath them when used as cladding materials. Viability in their size and shape, their high performance in diffuse light conditions, as well as the ability to obtain mechanical flexibility, are also important attributes that allow their application/integration even on curved roof surfaces or side surfaces of constructions. For this case, the development of specialized systems for different power-cultivation dual-land uses is highly favored. Moreover, the usage of abundant and low-toxicity materials, and their lower energy processing needs, promote sustainable solar cells manufacturing. All the aforementioned benefits concise that this group of solar cells would be the ideal PV devices for their implementation in agriculture. However, in this case, it is the higher LCOE that still obstructs their widespread adoption in agriculture and their high market penetration compared to the mature c-Si PVs, mainly arising from their lower performance and stability.

5.5 Outlook

When trying to describe the agrivoltaics approach, probably the term solar sharing is the most descriptive way. Sharing the solar resource to produce both electric power and food effectively at the same time is the main challenge to this strategic aim. On the other hand, it should be considered that each case is different and the design of the agrivoltaic system cannot always follow a standard approach. System adaptations to the local climate, crop type, land shape and power needs should be implemented. Up until today, a wide variety of PV technologies have been proposed and applied in agriculture, with the mature c-Si solar cells to be the dominant case, while high-tech PVs that offer high performance, transparency and wavelength-selective absorption being recognized in the last years as valuable alternative options. In practice, it is perceived that there is no one-size-fits-all answer on which PV technology is the ideal one for this niche application, while it is better to avoid banking on which technology will be dominant for agrivoltaics approaches in the future. The content of the present chapter may help to raise awareness of the possible advanced options to build a more efficient electricity-food dual-land use and set the base for creating novel ideas and more effective foundations in the field.

References

1. Gibboney CN (1949) The united nations scientific conference for the conservation and utilization of resources. Science (80-) 110:675–678
2. Trommsdorff M, Dhal IS, Özdemir ÖE, et al (2022) Agrivoltaics: solar power generation and food production. In: Solar energy advancements in agriculture and food production systems. Academic Press, pp 159–210
3. Brite Solar (2021) Brite Solar greenhouse presentation. https://www.britesolar.com/brite-solar-greenhouse-presentation/. Accessed 6 Jul 2023
4. Katsikogiannis OA, Ziar H, Isabella O (2022) Integration of bifacial photovoltaics in agrivoltaic systems: a synergistic design approach. Appl Energy 309:118475
5. Bellini E (2022) Optimal topology for agrivoltaics based on bifacial PV modules. In: PV Mag. https://www.pv-magazine.com/2022/01/21/optimal-topology-for-agrivoltaics-based-on-bifacial-pv-modules/. Accessed 14 May 2023
6. Marucci A, Cappuccini A (2016) Dynamic photovoltaic greenhouse: energy efficiency in clear sky conditions. Appl Energy 170:362–376
7. Moretti S, Marucci A (2019) A photovoltaic greenhouse with variable shading for the optimization of agricultural and energy production. Energies 12:2589
8. Papakonstantinou I, Portnoi M, Debije MG (2021) The hidden potential of luminescent solar concentrators. Adv Energy Mater 11:2002883
9. Weber WH, Lambe J (1976) Luminescent greenhouse collector for solar radiation. Appl Opt 15:2299–2300
10. Essahili O, Ouafi M, Moudam O (2022) Recent progress in organic luminescent solar concentrators for agrivoltaics: opportunities for rare-earth complexes. Sol Energy 245:58–66
11. (2017) Soliculture. http://www.soliculture.com/. Accessed 2 May 2023
12. Frias AR, Cardoso MA, Bastos ARN et al (2019) Transparent luminescent solar concentrators using Ln^{3+}-based ionosilicas towards photovoltaic windows. Energies 12:451
13. Purcell-Milton F, Gun'ko YK (2012) Quantum dots for luminescent solar concentrators. J Mater Chem 22:16687–16697
14. Bergren MR, Makarov NS, Ramasamy K et al (2018) High-performance $CuInS_2$ quantum dot laminated glass luminescent solar concentrators for windows. ACS Energy Lett 3:520–525
15. Raut HK, Ganesh VA, Nair AS, Ramakrishna S (2011) Anti-reflective coatings: a critical, in-depth review. Energy Environ Sci 4:3779–3804
16. Dey T, Naughton D (2019) Nano-porous sol-gel derived hydrophobic glass coating for increased light transmittance through greenhouse. Mater Res Bull 116:126–130
17. Ma Q, Zhang Y, Wu G et al (2023) Study on the effect of anti-reflection film on the spectral performance of the spectral splitting covering applied to greenhouse. Energy 272:127074
18. Yano A, Onoe M, Nakata J (2014) Prototype semi-transparent photovoltaic modules for greenhouse roof applications. Biosyst Eng 122:62–73
19. Cossu M, Yano A, Li Z et al (2016) Advances on the semi-transparent modules based on micro solar cells: First integration in a greenhouse system. Appl Energy 162:1042–1051
20. Wiesenfarth M, Philipps SP, Bett AW, et al (2017) Current status of concentrator photovoltaic (CPV) technology
21. Sato D, Yamada N (2020) Design and testing of highly transparent concentrator photovoltaic modules for efficient dual-land-use applications. Energy Sci Eng 8:779–788
22. Nardin G, Domínguez C, Aguilar ÁF et al (2021) Industrialization of hybrid Si/III–V and translucent planar micro-tracking modules. Prog Photovoltaics Res Appl 29:819–834
23. Shi H, Xia R, Zhang G et al (2019) Spectral engineering of semitransparent polymer solar cells for greenhouse applications. Adv Energy Mater 9:1803438
24. Chalkias DA, Charalampopoulos C, Andreopoulou AK et al (2021) Spectral engineering of semi-transparent dye-sensitized solar cells using new triphenylamine-based dyes and an iodine-free electrolyte for greenhouse-oriented applications. J Power Sources 496:229842
25. McCree KJ (1971) The action spectrum, absorptance and quantum yield of photosynthesis in crop plants. Agric Meteorol 9:191–216

26. Aira JR, Gallardo-Saavedra S, Eugenio-Gozalbo M et al (2021) Analysis of the viability of a photovoltaic greenhouse with semi-transparent amorphous silicon (a-Si) glass. Agronomy 11:1097
27. Dessì A, Chalkias DA, Bilancia S et al (2021) D-A-π-A organic dyes with tailored green light absorption for potential application in greenhouse-integrated dye-sensitized solar cells. Sustain Energy Fuels 5:1171–1183
28. Chalkias DA, Charalampopoulos C, Aivali S et al (2021) A di-carbazole-based dye as a potential sensitizer for greenhouse-integrated dye-sensitized solar cells. Energies 14:1159
29. (2017) Brite Solar in Ypaithros Hora. https://www.britesolar.com/brite-solar-ypaithros-hora-greenhouse-produces-energy/. Accessed 14 May 2023
30. Barichello J, Vesce L, Mariani P et al (2021) Stable semi-transparent dye-sensitized solar modules and panels for greenhouse application. Energies 14:6393
31. Ntinas GK, Kadoglidou K, Tsivelika N et al (2019) Performance and hydroponic tomato crop quality characteristics in a novel greenhouse using dye-sensitized solar cell technology for covering material. Horticulturae 5:42
32. Li Y, Zheng Y, Yan C et al (2023) Spectral engineering: the key to high-performance semitransparent organic photovoltaics. Sol RRL 7:2300037
33. Song W, Fanady B, Peng R et al (2020) Foldable semitransparent organic solar cells for photovoltaic and photosynthesis. Adv Energy Mater 10:2000136
34. Zhao L, Huang X, Wang Y et al (2023) Combination of highly photovoltaic and highly transparent materials enables record performance semitransparent organic solar cells. Chem Eng J 451:139081
35. Friman-Peretz M, Ozer S, Geoola F et al (2020) Microclimate and crop performance in a tunnel greenhouse shaded by organic photovoltaic modules—comparison with conventional shaded and unshaded tunnels. Biosyst Eng 197:12–31
36. Waller R, Kacira M, Magadley E et al (2022) Evaluating the performance of flexible, semi-transparent large-area organic photovoltaic arrays deployed on a greenhouse. AgriEngineering 4:969–992
37. Beattie D An aesthetic solar greenhouse. https://www.asca.com/projects/a-greenhouse-equipped-with-asca-solar-modules/. Accessed 17 Jul 2023
38. Safat Dipta S, Schoenlaub J, Habibur Rahaman M, Uddin A (2022) Estimating the potential for semitransparent organic solar cells in agrophotovoltaic greenhouses. Appl Energy 328:120208
39. Hollingsworth JA, Ravishankar E, O'Connor B et al (2020) Environmental and economic impacts of solar-powered integrated greenhouses. J Ind Ecol 24:234–247
40. Subhani WS, Wang K, Du M et al (2019) Anti-solvent engineering for efficient semitransparent CH_3NH_3PbBr 3 perovskite solar cells for greenhouse applications. J Energy Chem 34:12–19
41. Wang Z, Zhu X, Feng J et al (2021) Semitransparent flexible perovskite solar cells for potential greenhouse applications. Sol RRL 5:2100264
42. (2023) National renewable energy laboratory: photovoltaic research. https://www.nrel.gov/pv/. Accessed 11 Jun 2023
43. Fthenakis V, Lynn P (2018) Electricity from sunlight: photovoltaic-systems integration and sustainability, second. Wiley
44. Kost C, Shammugam S, Fluri V, et al (2021) Levelized cost of electricity renewable energy technologies
45. Trivedi H, Meshram A, Gupta R (2023) Recycling of photovoltaic modules for recovery and repurposing of materials. J Environ Chem Eng 11:109501
46. Heinrich M, Kuhn TE, Dimroth F, et al (2020) A comparison of different solar cell technologies for integrated photovoltaics. In: 37th European photovoltaic solar energy conference and exhibition

The manufacturer's authorised representative in the EU is Springer Nature Customer Service Centre GmbH, Europaplatz 3, 69115 Heidelberg, Germany. If you have any concerns regarding our products, please contact ProductSafety@springernature.com

Printed and bound by CPI Group (UK) Ltd, Croydon, CR0 4YY
01/10/2026
02225919-0009